What People Are Saying About
Anne Therese Gennari

"A wonderful definition of [...] all the voices. Anne Therese Genna[ri...] studied the science of global wa[rming...] Handbook lays out where we are, [...] forward. Fully grasping the implicatio[ns of the] climate emergency is not for the faint of heart. Understanding how to address the impending crisis will fill your heart. Here's how."
— Paul Hawken, Author of *Drawdown* and *Regeneration: Ending the Climate Crisis in One Generation*

"Do you want to help the planet and reverse global warming? Don't worry. You don't have to join a march. You don't have to shame people on social media or your best friend who wants to use a straw when you go out to eat. You can contribute in any number of ways and pick those that make you feel comfortable. Anne Therese Gennari offers practical methods for joining in the global effort. Start here if you want to feel empowered about our future."
— Per Espen Stoknes, Author of *What We Think About When We Try Not to Think About Global Warming and Tomorrow's Economy: A Guide to Creating Healthy Green Growth*

"This book teaches us how to move from a place of doom, gloom, and disaster to one of opportunity, curiosity, and hope. It reminds us of how good the world used to be and could be again and how even today we are making progress toward a better future. We have more to do, but a positive future is possible as Anne Therese Gennari shows."
— Ingmar Rentzhog, Founder of *We Don't Have Time*

"I always say ENVIRONMENTALISM + OPTIMIST + ACTION = INNOVATION & HOPE and that is exactly what Anne gives us with this book!"
— Alysia Reiner, Actress, Activist & Producer

"In the dark night of the climate collapse and emergency, *The Climate Optimist Handbook* is a ray of light and hope for all of us to read for inspiration and direction. Anne Therese has created a beautiful book for the moment and tomorrow."
— Paul Samuel Dolman, Host of the *What Matters Most Podcast*

"An oracle for younger generations that will inherit stewardship of this planet, Anne Therese Gennari's handbook will both educate and inspire, move and cause you to reflect, on how to become your most courageous and fulfilled self in relation to the greatest challenge humanity has ever faced. And if you're in doubt, read Chapter 3 right away!"
— Jack Adam Weber, Author of *Climate Cure, Heal Yourself to Heal the Planet*

"*The Climate Optimist Handbook* is a breath of fresh air, moving us away from doom scrolling to taking action, from worrying about our planet's future to instead building the world we need, one practical step at a time. Anne as a long-time Remake Ambassador and ally to our movement, exudes joy and hope as does her book."
— Ayesha Barenblat, Founder and CEO, *Remake*

"At a time when the headlines are filled with doom and scaremongering around the state of our world, I often look for an escape. However, the concepts of this work have helped me to explore and delve deeper within my own psyche for answers to find solutions in a positive and achievable way. Anne Therese Gennari has created a handbook that has me feeling uplifted and encouraged in the future of our planet, as it now becomes the responsibility of us, the climate optimists."
— Pauli Lovejoy, Drummer, Recording Artist, Music Director, and Model

"*The Climate Optimist Handbook* is a practical, realistic look at what we can do as individuals to reverse climate change. Some of us feel the situation is hopeless, and some of us would rather not think about it and mistakenly think it's something in the distant future, but Anne Therese Gennari shows that the situation is dire and needs addressing now. But she also shows why there is reason for hope if we just pull together and

continue to be creative and innovative like the human race has always been. Read this book and open your eyes to the better future that awaits us all by learning better ways to live—who doesn't want that?"
— Tyler R. Tichelaar, PhD and Award-Winning Author of *The Best Place* and *When Teddy Came to Town*

"If you're sick of doom scrolling through climate change headlines and are ready to roll up your sleeves for a more sustainable, equitable world, let Anne Therese Gennari be your guide. This persuasive, inspiring handbook flips the traditional climate narrative on its head and shows us how we can move past our fears with courage and conviction to create meaningful impact on the biggest challenge of our time."
— Andreas Karelas, Founder of RE-volv and author of *Climate Courage: How Tackling Climate Change Can Build Community, Transform the Economy, and Bridge the Political Divide in America*

"An easy-to-read, hopeful, and inspiring book. Anne does an amazing job at helping us embrace not only the challenge of change but the adventure of change."
— Dr. Chris Macdonald, Scientist and Author of *Operation Sustainable Human*

"As a storyteller, I'm slightly obsessed with narratives. But not all narratives. The climate narrative is one of those that urgently needs to change — from "the end of the world", to "a new beginning". Because you know what: We can still fix this. You don't believe me? Read this book. Then you will."
— Markus Lutteman, Bestselling Author and Climate Communicator

"*The Climate Optimist* is a critical call to action showing us how each of us can make a positive change in shaping the future of our planet. The path to a better future will always involve some degree or level of uncertainty. This powerful book offers encouragement, wisdom, and practical suggestions of simple things we can do now to make a difference for future generations."
— Susan Friedmann, CSP, International Bestselling Author of *Riches in Niches: How to Make it BIG in a small Market*

"Anne Therese sends us a much-needed reminder that the term 'empower yourself' is a silly phrase. You already have the power within you; you just need to use it. After reading her book, you'll want to more than ever."
— Rebecca Soffer, Author of *The Modern Loss Handbook: An Interactive Guide to Moving Through Grief and Building Your Resilience*

"The most important manifesto for our generation. Optimism is necessary for action. And it's optimists who will create the future. AT's message is not just a feel-good thing — it's the neuroscience of making things happen. Every climate activist needs to need read this book and practice its teachings every day."
— Katie Patrick, Author of *How to Save the World, How to Make Changing The World The Greatest Game We've Ever Played*

In *The Climate Optimist*, Anne Therese Gennari provides a thorough look at what we can do to reverse climate change and why we have reason for hope. As a behavioral scientist I know that it's critical we lead with hope if we are going to move people to change their behaviors. With hope and the tools we have cultivated globally, we can indeed overcome the climate crisis. This book will show you how to be most effective in your actions.
— Sweta Chakraborty, Climate Behavioral Scientist

THE
CLIMATE
OPTIMIST
HANDBOOK

How to
shift the narrative on
climate change and find
the courage to choose change

Anne Therese Gennari

The Climate Optimist Handbook:
How to Shift the Narrative on Climate Change and Find the Courage to Choose Change

Copyright © 2023 by Anne Therese Gennari. All rights reserved.

Published by:
Aviva Publishing
Lake Placid, NY
(518) 523-1320
www.AvivaPubs.com

All Rights Reserved. No part of this book may be used or reproduced in any manner whatsoever without the expressed written permission of the author. Address all inquiries to:
Anne Therese Gennari
www.TheClimateOptimist.com

ISBN (PAPER): 978-1-63618-206-3
ISBN (E-BOOK): 978-1-63618-244-5

Library of Congress Control Number: 2022917507

Editors: Tyler Tichelaar and Larry Alexander, Superior Book Productions
Cover Design: David Provolo
Interior Book Layout: David Provolo
Author Photo: Starky Morillo

Every attempt has been made to properly source all quotes.

Printed in the United States of America

First Edition

2 4 6 8 10 12

*To my daughter,
who will see the light of the world
at the same time as this book,
and who's already restored my faith
in our ability to get this right.*

CONTENTS

Introduction 14

PART ONE SHIFTING THE NARRATIVE 19
Chapter 1 Let's Not Be Chickens 20
Chapter 2 Choosing Change. 22
Chapter 3 From Angry Activist to Climate Optimist 27
Chapter 4 The Doom and Gloom Is Not Helping Us. 32
Chapter 5 It's Cheaper to Save the Planet 41
Chapter 6 It's Not Your Fault 45
Chapter 7 Do I Need to Raise Chickens? 50
Chapter 8 We Are All One. 54
Chapter 9 An Intersectional Matter. 56
Chapter 10 Leave a Big Footprint 60
Chapter 11 Capitalism—Good or Bad? 63
Chapter 12 Tone Down the Urgency 67
Chapter 13 You Are a Tiny Fish in a Huge Ocean 70
Chapter 14 Fueling the Transition 72
Chapter 15 From Responsibility to Opportunity 75

PART TWO CHOOSING CHANGE 77
Chapter 16 Choosing Change 78
Chapter 17 Generational Amnesia 80
Chapter 18 Two Different Futures 85

Chapter 19 Two Types of Change 90
Chapter 20 Curtains Can Ruin Christmas 95
Chapter 21 Fear of Change 99
Chapter 22 Complain a Little, Then Get Used to It 102
Chapter 23 Being Wrong 108
Chapter 24 Retruthing 113
Chapter 25 A Coded Computer 118
Chapter 26 Finding a New Truth 125
Chapter 27 Shame. 131
Chapter 28 Normal 135
Chapter 29 Four Teenagers and a Car 140
Chapter 30 Road Tripping. 147
Chapter 31 Keep Some Room in Your Heart for the Unimaginable 153
Chapter 32 A New Future. 159
Chapter 33 The Human Agenda. 162
Chapter 34 We're Going. 165

PART THREE AWARENESS AND HEALING 167
Chapter 35 Awareness Crashing In. 168
Chapter 36 Ignorance Is Not Always Bliss 171
Chapter 37 Awareness Hurts and That's Okay 176
Chapter 38 Awareness Overload 184
Chapter 39 Climate Anxiety. .188
Chapter 40 Activate Healing .193
Chapter 41 What Else Is a Heart For? 201

Chapter 42 Raising Your Fear Mark 204
Chapter 43 Dealing with Negative Climate News 209
Chapter 44 Should I Have Kids? 213
Chapter 45 Let Your Spirit Catch Up 216
Chapter 46 Choosing Light 220
Chapter 47 AT's Tips for Climate Healing. 223

PART FOUR CHOOSING OPTIMISM 229
Chapter 48 Optimism Is a Tricky Thing. 230
Chapter 49 Understanding Optimism. 234
Chapter 50 Three Reasons for Optimism. 238
Chapter 51 Anger 247
Chapter 52 How to Be an Optimist in Action. 252
Chapter 53 Your Secret Sauce 258
Chapter 54 Happiness Hormones. 261
Chapter 55 Optimism in Practice. 272
Chapter 56 Why Climate Scientists Are Optimistic 275
Chapter 57 Diamonds on the Soles of Her Shoes. 277
Chapter 58 Magdalena from the Future 280

PART FIVE CHOOSING EMPOWERMENT 285
Chapter 59 Activate. 287
Chapter 60 Planting Seeds 290
Chapter 61 Do Individual Actions Matter? 294
Chapter 62 Don't Be an Angry Vegan. 298

Chapter 63 Titles 301

Chapter 64 Successful Climate Communications 305

Chapter 65 *Lagom*—Not Necessarily Perfect 312

Chapter 66 Slow 314

Chapter 67 I Think You Should Hug Trees 318

Chapter 68 How to Think Like a Climate Optimist 320

Chapter 69 Leaders Create Leaders 323

Chapter 70 Scary AF 325

Chapter 71 Things Will Get Worse 327

Chapter 72 Paradise on Earth. 329

PART SIX CLIMATE OPTIMISM 323

Chapter 73 Things to Be Excited About. 333

Chapter 74 Inspiration for Individual Action 337

Chapter 75 Terminology. 357

Acknowledgments 374

Bibliography. 375

About the Author. 378

Bring Climate Optimism to Your Next Event. 379

It's time to shift the narrative on climate change so we can act from courage and excitement, not fear.

The climate crisis is not a climate problem, nor is it an environmental, political, or economical problem. It is a human problem. We are in this mess because of humans, and humans are the only ones who can get us out of it. Therefore, although this book is about a lot of things, and a great deal about climate, it's mainly about humans.

It's a book about you.

INTRODUCTION

Imagine you're standing at the edge of a cliff. On the horizon, a big, dark cloud is approaching, the early warnings of a storm. You're human but live a long time ago, ages before any civilization took root on Earth. Your home is a cave in the wall, and your survival depends on your ability to read your surroundings. You are at the mercy of nature, brought up to respect equally its ferocity and beauty.

What do you do? The Almighty has respectfully shown you what's about to come, so what will your actions be? If you had any plans to explore neighboring lands, they will have to wait. Every instinct in your body is telling you to retreat and stay safe for now. Maybe tomorrow, once the storm has passed, you'll muster the courage to venture out. For now, you want to get your family inside and make sure you have enough food and firewood to hold you over.

Sensibility. Safety. Survival. Pretty smart moves.

Today, we have a myriad of modern amenities that can keep us safe, even in a storm, but our deeper instinct for survival is still there. When we can't control what's about to happen, we're programmed to retreat and keep ourselves safe. When our environments seem unpredictable, messy, and unsafe, we don't choose change—we stay where we are, tightly holding on to what we know is working.

You're not as quick at quitting your job when the economy is unstable, and you won't necessarily leave your broken relationship when other things suddenly go off the rails. You may have thought about leaving for a long time, but a sudden death in the family leads you to different thoughts. If you wanted freedom and something new before, now turbulence in your safety circle can quickly cause you to grasp comfort and any signs of love, even if they are outdated and broken.

It's human to act this way—it's what we've been programmed to do for millennia—but what we don't see is that right now we need to let some of that sensibility go. We have to ignore our instinct, go against our nature, and dive headfirst into the storm. We must leave comfort behind when things are uncertain and trust that things can get better. Why? Because this storm is not passing by—it's stalling and only growing by the day, and if we hide for too long, the world might be so different when we finally crawl back out that we won't be able to live here at all.

Climate change is not like a comet speeding toward Earth. Climate change is a scary storm we can see on the horizon, but we may not yet understand how bad it's going to be. And that's not just a metaphor—today's storms are actually growing crazier and stronger each year. One of the side effects of climate change is that the jet streams (fast-flowing, narrow, meandering air currents in the atmosphere) are being disrupted, causing storms to intensify and hover over one place instead of moving on. This is why a rainstorm doesn't just stop by for a day or two anymore—it can stall and keep raining on the same place for days, even weeks, causing flooding so disastrous people die and others have to abandon their homes.

Climate change as a metaphorical storm is just like that. We can't predict it; we don't know how to prepare for its damage, and every part of our body is telling us to please stay safe and go inside. Don't take risks. Don't look for new worlds. Stay right where you are.

Maybe you've asked yourself why on Earth we aren't doing more about climate change. Perhaps this very question has started to keep you up at night, leaving you unable to sleep because all you can think about is how we're handing a broken world to our children.

First off, you're not alone. A 2020 study showed that 74 percent of Americans believe climate change is probably or definitely affecting their mental health. We're increasingly worried, stressed, and anxious about this growing crisis, and yet it is so difficult to act—why?

Well, think about the Stone Age. As humans, we're simply not programmed to expand our horizons when uncertainty is knocking on our door. And when it comes to climate change, uncertainty should be its middle name because everything we take for granted today is changing. The climate is changing, the weather is getting weirder, coastlines are shifting, and our thoughts of the future are shifting with them. We're in the midst of a lot of chaos right now, and as if that were not enough to manage, we keep fueling our anxiety with more of the same news. More fear-inducing articles about how the Earth is heading toward tipping points and how we're running out of time to do something about it.

For those of us who have the energy to care about climate change, all we hear is how we "need to act now or else we'll fail future generations" and it's "our responsibility to do the right thing." Big words to throw at someone who's looking a storm in the eye and whose evolutionary response is to retreat and wait for better days.

The question is, how do we act in the midst of all this chaos? How do we find the courage to leave safety behind and explore new worlds, new worlds that are not only out of the eye of the storm, but perhaps better and even more beautiful than anything we've ever seen to date?

To start, we have to recognize what a tremendous challenge change is for humanity so we can amp ourselves up for the task. If we think of ourselves as anything short of revolutionaries, we will fail.

To choose change when everything in our environment tells us to do anything but that is to be—as many would frame it—crazy. Yet we need crazy right now. We need people who have the courage to do things that have never been done before and who find hope in places that may appear to be nothing but dead-ends. Truly, we need a world full of heroes.

By the end of this book, I'm hoping you are both willing and eager to be one of them.

All it really comes down to is our relationship to change and how good we are at not only accepting it but finding ways to choose it. The one thing we know with absolute certainty is that things are changing. As Swedish environmental activist Greta Thunberg famously said, "Change is coming whether we like it or not."

But what if change isn't a bad thing? What if change is the best thing that could ever happen to us? What if we just have to wake up and see it? Maybe what it all comes down to is we have to choose change before change chooses us. And maybe, if we took a second to really think about it, that is, in fact, a very, very exciting thing.

We tend to worry because we don't know what the future will look like. We fear change because we don't know what that change will bring. But we also know that everything we've ever built, invented, or created has come from an ability to look beyond what we know now. Sparked by a curiosity to find something different, humans have birthed new ideas throughout history from crazy, wild, and courageous hearts.

Think about Thomas Edison who brought light into our homes in a new way or the Wright brothers who found a way to put humans into the sky. Before light bulbs or airplanes existed, flipping a switch to get light or jumping geographical borders probably seemed like "out of this world" ideas because they were.

But then we created a new world, and another new one, and another new one, and I bet you take it for granted that you can be across the country tomorrow. Think of all the things we take for granted today that, if we lived only a hundred years ago, would seem completely crazy.

Our world today consists of millions of things—big and small—made by people who had the courage to dream of something new. We have things made by people who were brave enough to think of realities that lingered outside their comfort zones, who dared to believe in the unbelievable and imagine the unimaginable, and who found the courage to go explore those new worlds. It's easy to forget that our reality today was sparked by crazy people from different worlds and different realities—dreamers who knew deep inside there must be something else. Crazy, courageous dreamers who envisioned better realities and wouldn't give up until they found a way. Yet that is exactly how it went down, and how it must go down again.

When it comes to climate change and all the other chaos we face at the moment, we have to tap back into that curiosity. We must expand our hearts and minds, question everything, and dare to dream. Because maybe change isn't so bad; maybe change means things can get even better—better in the most unimaginable ways.

There's a famous quote that's been circulating for decades and that has taken different forms over the years. I especially love how it was shared by the prominent sustainability architect and author William McDonough: "The Stone Age did not end because humans ran out of stone. It ended because it was time for a rethink about how we live."

We're there right now, at the bridge to a new era.

And you get to help write the script for what comes next. Are you ready? Are you ready to become a hero and tap into your own source of curiosity and courage? Are you prepared to believe, to fill your heart with hope and excitement, to join the journey to a different world—not lesser, harder, with more taxes or less freedom and control, just different, better, healthier, with safer jobs and saner hours, and with more time to spend with those we love?

It all starts with finding the courage to dive into those bigger questions. To ask ourselves, "What world do I want to believe in, and how can I play a role in making that world?"

The storm is approaching; are you ready to dive in?

PART ONE
SHIFTING THE NARRATIVE

CHAPTER 1
Let's Not Be Chickens

With climate change, we're usually asked to take urgent action by scientists and activists on the verge of panic. "World hurtling to climate danger zone, leaders 'lying' about their efforts." Or "It's now or never to fix climate change!" Google "climate change" to see what I mean. But the issue is that humans don't function that way. If you tell us it's urgent and the end is nigh, we'll panic. But then what? Urgency and panic are powerful energies that, when not properly managed, can do more harm than good. When you can't funnel that pent-up energy into something productive, chances are you'll get so overwhelmed by the whole thing you simply say, "Screw it" and shut down.

I like to picture us as chickens in a coop with a fox approaching. We panic and run around clucking, "What do we do? What do we do?" while creating a thumping dance of noise and chaos. Although a humorous scene, there isn't much productive action, and as soon as we realize there's no way of escaping—that we're stuck—what do we do?

If you're like most people, chances are you'll hide and hope you can stay hidden until it's all over. Ignorance is bliss—isn't that what they say?

When we really need to wake up and act, we are instead overwhelmed by fear and our inability to move leaders to act, so most of us are too afraid even to think about what's going on.

We need to figure out a different way. First, we have to get excited about the coming technological changes. To get excited, we must begin by telling ourselves things can (and will) get better. We must allow ourselves to dream of a world we've never seen and then wake up in the morning and somewhere, if only deep inside, believe it's possible. We must believe we can do this—that we have what it takes to create a world unlike anything we've ever seen.

This book is not about climate change; this book is about you. You are the vessel for change. You hold the power to say yes or no to a better world. You are the gateway to a better tomorrow.

At first, that might seem confusing and maybe even a bit overwhelming, but if you take a second to think about it, you might feel more grounded than you have in a while. We have no control over either climate or the weather. They are forces that should be both respected and held in awe, just as in the Stone Age. But we do have control over ourselves. We do hold the power to change our ways, and in doing so, change our path moving forward.

This is our time, our moment, our chance to make history and cocreate a better world! Let's heed our Mother's call and join her. Let's grow courage in our hearts and tell ourselves we can do this because we have what we need to leave panic behind in search of something better.

You will see the word "cocreate" a lot in this book. A few brilliant people may have made history in the past, but they didn't face what we're facing now. This challenge, this moment in time, is unlike anything we've ever seen. It's a challenge of unbelievable proportions, but it can turn into an opportunity of the same scale if we choose to act. For this task, we can't rely on solo heroes. For this challenge, we need people. People just like you and me who are flawed, innocent, scared, curious, and excited. People who have made mistakes and will continue to make mistakes, but in doing so, grow and become part of the evolution, part of the vehicle that will take us to a new world.

I know venturing into this storm is no small task and, yes, it will be scary at times. But would you rather wait at home and live through the disaster, or embark on an exciting adventure and lead us into something new? Since you're reading this book, I think we both already know your answer. And don't worry; I will do my best to guide you through the necessary steps to becoming a change maker. Let the following pages serve as your guide to becoming your best self so you can show up as the person our world so desperately needs you to be. You are the change. All you have to do is say yes to activate your power.

It starts with giving yourself permission to believe in a better world. The rest will simply follow.

CHAPTER 2
Choosing Change

One Friday night in May, I found myself in one of the most lavish apartments I've ever been in. It was one of those places with windows from floor to ceiling and big, heavy curtains caressing their frames. A huge wooden table was centered in the middle of the room, and in every corner, there was either a sculpture or an impressive painting. If you walked up to the window, you looked right out at the cobbled streets of lower Manhattan.

How I ended up at an intimate dinner party in Tribeca with a group of documentary producers, I didn't know, but it wasn't the first time my path had brought me to some cool places. Sometimes life throws you interesting curveballs, and I've swung at enough of them to know how to sip cocktails with magic when it chooses to appear. You simply don't know what will come from saying "yes" to an invitation, so when asked if I was in New York and interested in joining a group of climate nerds for dinner, I decided to accept.

"How do you think your perspective as a Swede differs from that of most Americans when it comes to dealing with the climate crisis?" I was asked.

I had just made the comment, "Of course, the movie would have a happy ending—it's an American movie after all," while we were bouncing around ideas for a blockbuster movie based on real-time and all-too-true climate facts. When the "happy ending" emerged, I had to say it—Americans.

Everyone laughed, and the joke spun on how the film would end if it were made in a country like Sweden. Would we all eventually die? And would that be the more realistic ending to the movie called *Climate Change*? (Note, this conversation took place before *Don't Look Up* reached theaters.)

With all six pairs of eyes now looking at me over the candlelit table,

I sobered up from my joke. I wanted to give them a sincere and thought-through answer, so I did a deep dive into my cultural background. What about me and my view now is shaped by American culture, and what have I brought with me from my upbringing in a Nordic country like Sweden? What is new, what is old, and what is just…me?

That was the first time I honestly reflected on my unique situation as a Swede in the States.

I've never been like most Swedes. I know that because I've never been much like anyone I know. But naturally, my upbringing in Sweden made a difference in who I am today, and surely my love for America has left a footprint as well. I am undoubtedly influenced by my roots, and I can relate to Swedish environmental activist Greta Thunberg and how she thinks the world is a messed-up place, yet I'm infused with the American Dream that lured me and so many before me to this soil.

I do believe climate change is a crisis we ought to treat with full respect and urgency, but I also appreciate the heroic spirit that fuels this country, the hope, wishes, and courage that make up the American Dream. It's the spirit of "Yes we can" and "I have a dream"—the kind of will that got us to the moon and makes people of all ages leave their hometowns to seek fame in the Hollywood Hills. It was this promise of a better life that first got me to leave home in my late teens to seek the American adventure—the possibility of finding an answer to the question: "What if?"

I gave the party what I believe was a satisfactory response, and the conversation went on to deeper levels of culture, politics, and social influence. However, I couldn't shake the question, and it followed me to the subway and home. What will it actually take to *fix* this? Considering that climate change is such a perplexing problem that no one solution can fix it and no one country has the power to change its course, how will we make it out on the other side?

For one, it will take all of us. Not just every country, but we, the people, need to trust in leaders' bold actions and look around ourselves to see how we can help. That means we must be willing to take a serious look at the future we're headed for and ask ourselves if that is what we want, and if it isn't, we must be ready to change our ways. Quite obvious, perhaps, but here comes the tricky part—our relationship with both the future and change!

As a species, we're infamously bad at thinking about the future. Most of us can't even save enough money to retire or realize that how we're treating our bodies now will not serve us well when we get older, so how do we start thinking about our future selves? Or even more so, how do we think about times when we will no longer be around? How often do we think about life on Earth after we are gone, when even our kids and grandkids have grown old and passed on, and the world will probably be very different than it is today? My guess is not very often.

However, there will be generations after us. You and I both know this, and they will want to thrive on this planet as we have. To be a good citizen means to recognize that, and when we do, we soon realize the truth of my next point—the simple fact that we must begin to choose change. For if we want those future unknown people to have a chance at life, we need to change. We're simply not giving them a great chance.

> *"Climate change is first and foremost a problem of our relationship with the future."*
> — Alex Steffen, as quoted in The Future Earth by Eric Holthaus

Let me ask you a question: When was the last time you did something bold? Maybe you moved to a new town, switched careers, or decided to finally leave a relationship that wasn't working. And if you're too young for any of that, maybe you finally decided to speak up for yourself or stopped hanging out with a friend who didn't make you feel good. Whatever it was, reflect on what brought you to take that action and how it made you feel. Looking back, do you wish you would've found the courage to change things for the better sooner?

Humans are an interesting species. We have this ability to make calculated choices and know what's ultimately best for us, yet it can be so hard to act on that knowledge and change things for the better. Our human relationship to change and our ability to actively choose (or not choose) to change has always fascinated me. What does it take to leave what we know is safe, the "good enough," in search of new opportunity, expansion, and growth? And more importantly, what triggers this kind of courage, and how can we use it to fuel the climate action we so desperately need?

The world needs change, and for that, we need courage, so the answer to the question I was asked over the candlelit dinner and that kept living inside me for quite some time, is we need it all. We need my perspective—new and old—and we need yours. We need the perspective of scientists, engineers, and economists, but also the perspective of the Indigenous and the old. We need the perspective of dreamers and visionaries, of artists and makers, of mothers and fathers, children, blue-collar workers, and entrepreneurs. We. Need. It. All.

As much as we need to focus on science and finding the smartest and fastest next step, we also need a little dose of crazy. We need wild hearts dreaming of seemingly impossible tomorrows, and we must nurture those hearts as often as we can. Because the truth is that so much is being asked of us right now. Not only do we need to look the storm in the eye and still choose to venture into it, but we also have to make amends for our mistakes and tell ourselves we can do better—that even with the good things, we need to look for change.

We have made it to the moon, to Mars, and to the deepest parts of our seas. Now we have to find what it takes to get to the regenerative, just, and climate-neutral future we all deserve. It may not seem like a big deal when you compare it to reaching the moon, but trust me, the challenges we're facing are so much bigger than anything we've faced before.

You may not know exactly what we face or what we need to do yet—I don't think any of us really does. We don't know what awaits us or the strength it'll require from our hearts to overcome it. The best thing we can do is prepare ourselves and recognize that we don't have a second to lose. All we have is something better to win.

That is why we need it all. We need fear and worry so we can remember why choosing change is necessary in the first place, but then we also need curiosity, excitement, and joy so we can find the motivation to keep trying. We need to dream big and act boldly and continue to grow and nurture our hearts. We have to choose change, again and again and again, until it becomes so natural that it's simply what we do. We say yes to change.

We have to question even when we don't think there's anything to question, and we must be willing to try new things, even if we're happy and comfortable with the old. And more than anything, we have to grow our relationship with the future and recognize that it's not about what we can accomplish, but how much we can nurture what we have and pass it on. It's not a rat race anymore; it's a journey, and one all of us get to travel together. That is why, although it might seem like a scary time to be alive, it's also the most exciting time there's ever been. It's a time when we get to do the heroic work of choosing change.

CHAPTER 3
From Angry Activist to Climate Optimist

The night it all changed, I had just had an argument with my brother about washing his car at home. We were having dinner with our parents, and per usual (because I couldn't help myself), I stirred up a conversation about the environment. This time, I made a fuss about how irresponsible it is to wash your car at home when there are car-washing stations. I felt specifically triggered because my brother didn't seem to share my worries.

"Don't you know that you're dumping a bunch of toxic chemicals right into the ground?"

I adore and respect my younger brother more than I do most others in this world. He has one of the biggest hearts of all the people I know, and he is smart far beyond his years. However, as my brother, he can be annoying too, and he loves to argue with his sister.

"Sure, sis, but come on; it's just one car. Do you think it really matters that much? I love washing my car at home; why should I give that up?"

This kind of discussion was a normal affair in our family, as I'm sure is the case in most households. Besides, this was before Greta Thunberg and before COVID-19, so no one was really freaking out about climate change yet. Besides me, of course. I was constantly freaking out about it, and I always felt desperate in my failed attempts to get others to care. This was nothing new. As the angry activist, I was used to starting a fight and losing the battle. However, this night, my brother pushed me over the edge. I don't know what it was exactly that cracked my system, and I've asked myself so many times why it was this conversation out of so many, but I guess my body had simply had enough.

My brother was making fair points. Sure, compared to so many industries around the world spewing toxins into the land, air, and sea every minute, a little at-home car washing wasn't going to tip the scale. And it wasn't like he was being a jerk, either. He was just defending something he loves, so of course, I knew he wasn't the evil one. I understood deep down the issue was so much bigger and that my brother was ultimately on my side. But I didn't hear any of that. What I heard was no one else cares, no one else gets it, and fighting climate change and saving the planet is a pointless, tiresome battle we'll never win.

I felt tears build up behind my eyes, and as I sat there at the dinner table, the people I love the most by my side, I felt so incredibly alone.

• • •

Knowing I was close to breaking down, I needed to find a place to be alone, so I brought my plate to the kitchen and walked to my parents' guest room. As soon as I closed the door behind me, I collapsed onto the floor, and in an instant, I started to cry. I have cried over many boys in my life, and I know what being sad, upset, and heartbroken feels like, so when I say I never cried like this, I mean it. It felt like the pain of the whole world, pain I must have kept bottled up inside me through years of anger and worry—all came flooding out.

It was weird. Like, out-of-my body weird. A pulsating force started in my belly and slowly spread through my upper body. My arms were bent in front of me with my palms facing up, completely locked in place. As I sat there and observed myself, I felt an energetic force beginning in my belly and exiting through my cupped hands. It was strange, but the experience was real, and I still remember it vividly.

I looked down at my paralyzed hands. I knew they were mine, of course, but I couldn't feel them. When I tried to wriggle my fingers, they wouldn't move. It was the strangest thing I had ever experienced, and I haven't had a moment like it since. I was mesmerized and confused at the same time. I couldn't help but wonder what was going on.

But then I understood exactly what was happening and decided to surrender. It was clear my body (and maybe even some higher force) had said, "Enough is enough!" It was time to let go.

I let go. I sat there on the floor and let whatever this was do its thing.

All I could think (because observing the scene from the third-person perspective gave me time to think) was I must've stored all this fear, pain, and grief up over the years, pain I never fully understood I was holding on to, and now I had reached a point where I couldn't hold it any longer. All the anger toward myself and the world, all the worry about what I can't control and frustration over the fact that I can't control it, had to be set free.

I felt emotionally moved by the love of these unseen forces. These higher powers—maybe my wiser self, a spiritual twin, angels, or God—who decided to step in and act changed my life. For so many years, I had fought to keep it all together, and I would've never given myself the right to collapse like this. But what began to dawn on me as I sat there was maybe all I could do was lose my tight grip and finally let go.

After what seemed like forever, the pulsating stopped. My eyes dried up as the crying ceased, and suddenly, I felt so…light. In a strange way, it felt as if I were floating above the ground. My arms, which had been locked in the same position the whole time, were tired, numbness in my muscles making them appear as if they were floating in place. But it wasn't just a physical feeling. Where there used to be a heavy rock inside my chest, there was now an expansive space, something completely different, a feeling I hadn't felt in quite some time.

Was it joy?

Not ready to stand up quite yet, I closed my eyes once more and tapped into the moment. As soon as I did, a beautiful white mist surrounded me. All I could see was a light fog and a magical, dim, white light. It's one of the most peaceful moments I've ever experienced, like one of those movie scenes where the character walks into the abyss and disappears into a cloud of white. I was in that nothingness and, for a moment, a part of me never wanted to go back.

Then a message came through. I'm not particularly religious, but I do believe something is out there trying to help us find our way. If this message came from one of those sources, I don't know, but I was cracked open and ready to receive it—finally able to listen.

Regardless of the messenger, the memo changed my life.

"You are here to be a climate optimist!"

What did it mean? I realized it didn't matter because receiving the message made me feel alive. For the first time in my young adult life, I felt like I didn't have to hide anymore, like I didn't have to choose between listening to my inner calling and living a good life. I felt like I was given permission to smile again, and although I had no freaking idea what being a climate optimist meant, I knew I was going to embrace that mission with passion.

Ever since that day, I have continued to seek answers about my mission. I refer to that day, this strange and somewhat otherworldly experience, as my climate optimist awakening. If you hear me say that—in this book or elsewhere—now you know what I'm talking about. I've let you in on my secret and shared what's probably the most non-sciencey stuff you'll read in this book. But I remember it clearly. I know it happened, and now you know too.

I know in my gut that I had that moment for a reason. My hope is it can serve as an awakening for us all. Maybe if I can teach you all the things I have learned and figured out since that day, you won't need otherworldly intervention before living life with purpose, passion, and excitement. Maybe I can help you activate that journey right now.

I didn't choose this path; it was chosen for me. Whenever I doubt and start questioning myself, I come back to that moment. I remember the message and how true it felt when it hit my heart. It took me many years and even more detours to find a way to make climate optimism a true part of my being, but now that I'm here, I never want to leave.

Climate optimism is nothing but a powerful mindset shift, a new look at yourself, the world, and the possibilities we hold moving forward. But the best thing about being a climate optimist (the way I'm about to teach you) is it will completely transform your life.

My aim with this book is to bring you in on my mission. I will spill all the secrets of what being a climate optimist means and give you the tools to become an optimistic, resilient, hopeful, and self-empowered person who lives to see the days beyond climate change. I can't give you superpowers (they don't exist), but that's okay. We don't need supermen and superwomen to save us. We need us. We need flawed, beautiful human beings who are willing to seek answers and muster the courage to choose change.

That is our calling. I know it in my heart, and by the end of this book, I hope you won't just be able to see that, but will be eagerly excited to get started.

CHAPTER 4
The Doom and Gloom Is Not Helping Us

The biggest narrative shift we have to make is on how we allow climate change to exist in our brains and in our culture. So far, we've associated climate change with doom, gloom, and disaster. We keep saying we need to act before it's too late, and we must step up to our responsibility and save the planet for future generations. The reason for action is always the same—"This is not what you want!"

Although it's true we don't want to continue on our current path to a disastrous climate future, how we present this dilemma many times inhibits action. It might seem controversial to say, but when we're faced with an overwhelming threat (like climate change definitely is), we actually feel less motivated to act. It goes way back to when our brains were wired for survival. When we approached a dangerous animal, let's say a tiger, we didn't have many options. If we were to try to fight the tiger, it would probably chew us to pieces, and trying to outrun it wasn't an option either. Therefore, to survive the encounter with a tiger, we did nothing at all. Our brain is wired to, in overwhelmingly scary situations like that, do as little as humanly possible. If we could only stay absolutely still and quiet, the tiger might not notice us and simply move on.

This is why fear can easily paralyze you, leaving you stuck in the moment. Your brain and body simply don't want to act. They think it's necessary to paralyze you to stay alive. Think about this the next time you're facing a big task and you simply can't get yourself to start. You know you have to, but the mountain of work ahead is so overwhelming that you simply can't muster the motivation to get going.

It's not your fault. Your brain is simply wired for survival, so give yourself a break.

As Tali Sharot, neuroscientist and professor of cognitive neuroscience at University College London and MIT, explains in *The Influential Mind*, our brain is wired to: A) respond to positive information and the potential of achievement with a "go response," and B) respond to negative information and the potential of a loss with a "no-go response." In simple terms, you will run toward the lush berry bush (or in today's world, the delicious-looking dish on a menu) because it means reward, and simultaneously, you will avoid anything that has to do with pain, sacrifice, or loss. Seems like a very smart way to live life; however, how often do we try to do the opposite? How often do we try to spark action—in ourselves and in others—by flagging the very threat we're trying to avoid? The "act now or else" mentality you think ought to work more often than not leads to complacency instead.

This simple psychology lesson is essential when it comes to climate change and the work we're trying to do. Naturally, our brains are more complex than that and nothing is ever as black and white as we try to make it. But for simplicity's sake, let's leave it at the "go" and "no-go" responses and have a look at how we usually communicate climate change:

"UN: World hurtling to climate danger zone, leaders 'lying' about efforts."

"UN scientists: It's 'now or never' to fix climate change."

"How chemical pollution is suffocating the sea."

"New temperature record—again!"

"Rapid glacier melt…."

"Climate tipping points…."

"Heatwaves…flooding…migration…sea level rise…extinction…."

These are just a few examples of climate change making headlines, but they are emblematic of how climate change is presented. The headlines are infused with disaster, fear, and alarming information we hope will push people to act when, instead, the opposite often happens.

If you've been around for a while, you might recall that climate communication didn't start out this way. When the news of a changing climate first began to reach the media, the headlines were much more subtle and infused with opinions and confusion.

"Space satellites show new Ice Age coming fast," was one headline in *The Guardian* in 1971. The article shared the belief that the continuous added "dust" in the atmosphere would lower temperatures and trigger a new ice age. Scientists were still confused as to what might happen—were we cooling or rather heating the Earth?

There was also deliberate doubt planted by Exxon Mobile, with headlines like "Climate change: a degree of uncertainty," that undeniably made people confused about whether any dangerous changes were going on, and if so, whether they were severe enough to sacrifice what we love—like driving! With so much uncertainty involved, one shouldn't be surprised that the incentive to act was weak, but how about now? Certainly, shouldn't more aggressive headlines trigger more action?

Ironically, that is not the case. In fact, the opposite might happen.

In *What We Think About When We Try Not to Think About Global Warming*, Per Espen Stoknes reports studies showing the more people learn about climate change, the less they tend to act to limit it. He calls it the "snooze button effect" where an alarm simply appears less and less intrusive when you've heard it two, three, four, maybe even five times. The first alarm will snap you out of sleep, but when you hit snooze once, and then twice, your body gets used to the sound, and you can almost keep hitting snooze for hours.

The same happens in your body with big, alarming threats. The first time you hear about climate change, you might feel the hair rise on your arms and have a very uncomfortable feeling spread through your body. You are alerted, frightened, and ready to fight, flee, or freeze. But the second time, and the third and fourth, you get more and more used to hearing about it. When you've heard it enough, it will simply exist as another uncomfortable buzz in the background you know you should do something about, yet lack the motivation to do so.

Obviously, this natural human reaction doesn't mean we should stop talking about climate change. As a matter of fact, we should talk about it more. But *how* we talk about climate change, and more importantly, the actions needed to change our direction is incredibly important. At the end of this chapter, I'll give some examples of how to shift the narrative and then look at changing the narrative in more depth later in the book. But before we get there, I want to share what Per Espen Stoknes calls the Five Ds (doom, distance, dissonance, denial, identity), also known as the five reasons we aren't doing more about climate change.

Doom

The first D is doom. Reports on climate change are filled with doom and gloom. If we read about it on social media or hear about it on the news, the headlines are rarely positive and, more often than not, quite scary. We also hear that we need to act now before it's too late and irreversible damage is near. Although these messages are true, how they're being presented triggers our freeze response—as with the tiger, neither the fight nor flee responses are viable options in our minds.

As you already know, negative information like this triggers a no-go response, a feeling of being overwhelmed, and a lack of incentive to do anything about it. This is why the doom and gloom messaging about climate change might be accurately alarming, but lacks the needed inspiration and motivation to create change.

Distance

The second D stands for distance. People need to relate to the concern to act. Whatever it is must touch us on a primitive level. Like most species, we evolved to care for ourselves and our kin first. It's simply an act of survival. If we don't feel the threat is imminent and urgent to our and our family's survival, it is incredibly hard to get motivated to act.

When it comes to climate change, it still feels distant to most of us. As a threat, it's perceived as something that's happening to other people in other parts of the world, to other species, or something that will happen in the future. And although we care about other people, animals, and future generations, the sense of personal urgency is not there, hence the difficulty in triggering action.

It does not mean we're mean, uncaring people. It simply means by nature we are not designed to act on distant threats, especially when we have so much else to worry about in our immediate surroundings.

Dissonance

The third D is (cognitive) dissonance. This is a conflict between our beliefs and actions. Let's say we worry about climate change and want to stop adding carbon dioxide to the atmosphere by burning fossil fuels, but the only way to get to work or to school is by driving. We may want to change, but society is simply not built to support us in that mission.

That is when cognitive dissonance kicks in, and it can be a tricky landscape to navigate. Do you keep hating yourself for contributing to the disaster? Do you give up life completely and start living off the land? Or do you start bending your beliefs and views just a little to better fit your current lifestyle?

For most of us, the latter happens. In order to function properly—both physically and mentally—we start making excuses for our habits so we can feel better about who we are. If the system is broken, how can we do anything about it?

Although I'm not saying cognitive dissonance is all bad, it's important to keep it from leading to complacency. We can recognize what isn't working and still choose to act. More on that later.

Denial

The fourth D is denial. Now, this one may trigger us because we're taught to believe denial is bad. Although it can appear ignorant and even cruel on the surface, denial is actually an act of self-defense. Just as it's impossible to live perfectly sustainable lives in our current world, it's also impossible to remain aware of everything that's going on constantly. If we did, we would soon lose ourselves and become rather crazy, with little left in us to inspire action.

Recognize that a bit of healthy denial is necessary. I will share tips on how to practice this later. But yes, denial is a huge roadblock to taking action against climate change. Acceptable or not, we need to find ways to step out of denial and into empowered action. And that, my friend, is why you're reading this book.

iDentity

The last and fifth D stands for identity. What does identity have to do with climate action? Everything. Stoknes discovered our social identity, the clues society sends us, are a huge hindrance to taking more action on climate change. We are social creatures and, therefore, constantly looking in our surroundings for clues as to what is okay, acceptable, normal, and cool. If everyone around us is still driving to work and eating meat, why wouldn't we?

To be the oddball and do something differently is scary at first, but it is also where you hold incredible power. If you start acting to bring about the world you wish to see, other people will soon recognize that, and a ripple effect will follow.

The more people do something new, be it bringing their own coffee cup to the store or switching to an electric car, the faster they form a new norm, and that is how we shift culture. We allow new identities to unfold and people to accept, trust, and build a new world.

Be the one who places solar panels on your roof, puts composts in your backyard, or creates a recycling program in your office. Be the one who takes the bike to work and wears the same outfit to many parties in a row, showing that it's cool to care and be mindful. The sooner we start acting on our beliefs and showcasing new ways of living with joy and excitement, the sooner we can kickstart the movement to build a whole different kind of world.

Are you starting to see why we aren't doing more about climate change and why we didn't do more in the past? Unfortunately, there is another huge factor—some don't care and some deliberately spread misinformation and bury or obscure evidence of our warming world. It's sad to know many people with power and money knew about and could have haltered this disaster decades ago simply by doing the right thing.

We can't change the past or curb today's greed and/or willful ignorance, so I'll leave that alone and bring the focus back to us. How can we get around the Five Ds, these five barriers, and ensure we're not just aware of the problem, but inspired to make a difference? The answer lies in shifting the narrative, changing the way we communicate about climate change, and most importantly, why and how we ought to act.

If it wasn't bad enough that headlines are infusing us with enough fear to paralyze us for a lifetime, how we talk about what needs to be done isn't much better. We are told to fix the problem by:

- Driving less.
- Eating less meat.
- Traveling less.
- Cutting down on consumption.

- Saying no to plastic.
- Eliminating food waste.
- Minimizing our footprint.
- Giving up consumerism altogether and living in the woods….

Although all these actions are accurate and needed to curb climate change, how they're presented does not motivate us. If we pay close attention to the message behind these actions, they all present the potential of a loss. Therefore, though they may be reasonable actions that would actually lead to better lives, our brain does not hear them that way. It hears we have to sacrifice things we love, remove some current pleasures, and ultimately, face losing much of what we love to do. What our primitive brain says to that kind of information is simply, "Thank you, but no thank you. I think I'm good as I am."

We need to start working with our brain, not against it, and recognize there are simple ways of saying the exact same things, just in different words. Here are some examples of how to shift the narrative to communicate the same actions:

- Let's vacation close to home so we can spend more time with each other, without the stress of travel.
- Let's try some plant-based foods. I've seen some recipes that look really good.
- We should switch to renewable energy to save money.
- I love my clothes; I should wear them more often.
- The world would be much more beautiful with cleaner air.
- Let's have deep conversations and connect.
- Look at the cool things we can do right here in our community.
- Locally grown food supports our neighbors.
- Mindfulness brings peace.
- Less running around means more downtime and moments for simply being.
- More…. (You fill in the blank)

If we pay attention, we see climate change is more than a huge responsibility; it's also a beautiful opportunity to rethink our world and how we live in it. We have an opportunity to look at everything and figure out what's not working and find even better ways. If we shift the narrative, we may begin to recognize that climate change can be a gateway into a new way of being, a renewed consciousness that brings us back to ourselves, each other, and the Earth. It doesn't mean we have to sacrifice and lose. On the contrary, we have an opportunity to grow and seek even more—more abundance, love, time, happiness, and wellbeing.

It's okay to be worried. I worry a lot too, and there are days when I also wonder if we even have a chance. To be a climate optimist is not to live in an endless cloud of light and wonder. You may lose hope sometimes, have days when you feel frustrated with yourself and the world, and days when all you want to do is give up.

Feeling that way is normal; it's human. No one is asking you to become superhuman. In fact, I believe we have to become as human as we can possibly be. That means we will have good days and bad days. Days when we feel inspired, optimistic, and strong, and days when we don't feel so high and just want to pull a blanket over ourselves and watch TV for hours. It's okay to have those days too. It's the human thing to do.

An extraordinary thing about being human is you have days when you are motivated to work for something better, days when you envision something even better than the world we know today and find the courage to see if that new reality could potentially come to pass. This ability to think of something that hasn't happened yet, to believe in a reality you haven't yet seen, is an extraordinary gift, and one you should honor and care for deeply.

To be a climate optimist means nurturing those qualities as much as you can. You recognize what an incredibly exciting time it is to be alive because we all get to be part of making history, do something that has never been done before, and come together as a whole to cocreate an even better world.

We tend to believe we have to sacrifice what we love to pay for our mistakes. But maybe we're blind in that belief. Maybe we think that way because we've never seen or experienced anything better. A world more beautiful and abundant is just a step away.

Some say it's too late. Others urge us to tap into our dormant anxiety and take action. I say we're right in the sweet spot where the answer is simple—don't act and the future will be freakishly scary, or find the courage to do what we can so whatever comes next can only bring wins. A better world is possible, and now is an incredibly exciting time to be alive. It means *you* get to be part of making history. It means you get to observe and participate in the shift toward a better world.

CHAPTER 5
It's Cheaper to Save the Planet

As a kid, I was always looking for new songs to play in my Sony Walkman when I biked around town. I spent a lot of my time glued to my dad's stereo in our living room waiting for the perfect time to hit record when the disc jockey announced that a song I liked was up next. Recording music from the radio was tricky because you had to be sure the tape cued up correctly and you pressed record just when the song started. Then you had to wait for it to be over so you could press stop. Often the DJ kept talking over the first notes, and you either got a bit of nonsense with your recording or a song that started a few seconds in. But it was the only way I could acquire new music, so I took what I could get.

I share this story because I'm not old. Anyone born after the millennia might think this scene belonged in the sixties, but even my parents were just kids then. I'm only in my early thirties, and this exciting but time-consuming hunt for new music took place in the late nineties. This was my reality growing up. Only twenty-some years ago, we had vinyl records, reel-to-reel tape, and cassettes, and compact discs were still new. If we wanted to record our own music, we used a cassette tape. The internet was only accessible after a long dial-up tone followed by extremely slow download speeds. If you don't know what a dial-up internet tone sounds like, you missed some cool times. Just saying.

One day when my brother and I were still young, our dad, who worked for a Swedish telecom company, came home and said he wanted to show us a video of the future.

"Come, kids; I want to show you something cool!"

His excitement was hard to hide, and our curiosity even more difficult to contain, so our family of four curled up on the couch in anticipation of seeing the "future world." And there we sat in disbelief as we watched a twenty-minute video about future technology, our mouths growing wider with each invention introduced on the screen. Would this actually come true?

One segment showed a man driving to a rock-climbing session and a computer in his car was giving him directions. In the next scene, a couple at the theater was worried about their dog at home. But they needn't worry, because, with a quick click on a device, they could see their pet on a tiny screen, knowing it was soundly asleep in its bed. They noticed the food bowl was empty, and with another click of a button, a pipe on the wall released food in the perfect serving into the bowl. As the dried pieces scrambled into the ceramic bowl, the dog sprang to life and ran over to the bowl, wagging its tail excitedly. The couple looked at each other and smiled.

We saw a student taking cello lessons with her teacher on a big screen in front of her, video calls between family members, and other technological advancements we had never laid eyes on before. We couldn't believe talking cars, clear video calls, and controlling your home with the push of a button could ever be true—it honestly seemed like it was out of this world.

At the time, these advances were out of that world. But we've created many new worlds since then, and our reality today includes GPS, video calls, and smart homes. It's easy to forget how far we've come, easy to miss how fast society has progressed over the last twenty years. Sometimes I have to remind myself of the five- to ten-minute wait for an internet connection. And the thirty-minute limit Dad gave us daily before we had to shut it back down. Think of that when you complain about a poor Wi-Fi connection underground.

The truth is humans are capable of extraordinary things, so when I say we can create a world where the air is so pure no child on the planet will have to worry, or where cities are among the healthiest places to live, you can believe me. When I say a time will come when cars are silent except the sound they emit to warn visually-challenged pedestrians and plants will hang from every building, can you see it?

I know it doesn't seem to make much sense in this world, but we're not staying in this world, are we? We're leaving it behind so we can learn from our mistakes and create much better worlds. As with any historic moment, our transformation will seem wildly impossible until it's done.

But that doesn't mean we can't make it. What it means is we have to continue tapping into our brilliant human brains and decide to create something new.

And that is why we're here, to cocreate a better world, and I want you to recognize the magnitude of the times we live in. This moment in history could be the most powerful of all time for homo sapiens on Planet Earth.

Will we step up to the challenge? Will we do the seemingly impossible and create something so completely new our mind's eye wouldn't recognize it if we saw it? That, my friend, is what we must begin to explore.

The single most dangerous stumbling block on this mission is our inability to think of and believe in new worlds. But once we give our imagination room to roar, magic happens. That is exactly what we'll get into in this book—how to imagine the unimaginable, get comfortable with change, and set some freaking magic in motion!

You don't have to be a climate nerd to be part of this change squad. If baseball and hotties are more your game, so be it. But do you want to watch baseball and hit on attractive singles in the future, or be stuck inside because the air is too toxic to breathe?

Climate change is already affecting us all, just not equally. If you're not a farmer in Africa or a family in an island nation in the Pacific, maybe you haven't been that concerned about climate change yet. But the truth is, unwanted change is upon all of us, and we must wake up from our slumber and see it for what it is. The question is actually quite simple—do we want a livable, healthy environment in the future, or do we not? Climate nerd or not, the answer should be easy.

The good thing is we don't have to sacrifice anything to build this new world. If that's what your belief system is telling you, you're tuning in to old news! To quote my friend and fellow climate optimist, Andreas Karelas, who said it so brilliantly on my podcast:

> It's actually cheaper now to save the planet than it is to destroy it. Meaning, it's more expensive to destroy the planet than it is to save it, so [climate change] is no longer an inconvenient truth [referring to Al Gore's film *An Inconvenient Truth*]; it's a convenient truth!

What does that mean? It means our technology and economy have made using the planet as a garbage dump more expensive than treating it with the care it deserves. It means that solar panels are now 99 percent cheaper than they were four decades ago and generate 121 times more power than the goal established in 2000.[1]

It means oil prices are going up and alternative solutions are more scalable than ever. It also means today's consumers want to buy from companies that genuinely care. The times when profits and shareholder value trumped corporate citizenship are waning.

The world is ready for change. It's already happening all around you, faster than you might understand. You, my dear friend, get to live in the midst of all this. You get to witness the world going from one reality to another, from one era to a new one, from one time to the next.

I want you to use these times as a beautiful opportunity to thrive, to expand yourself, your mind, and your beliefs in a better world, and in doing so, create a life for yourself more meaningful and influential than you could have ever dreamed of.

That is what our time here is all about—to expand, to transform, to thrive! That is what my climate optimist heart is telling me, and so it must be. Do you believe it too?

Get ready, children of our times, for the most significant transformation we've ever seen. The biggest shift the world has ever seen. I hope you're as excited as I am because we have no time to wait, so let's get to it.

Let's change the world.

---

[1] Al Gore, As heard in a Climate Reality Project Leadership Training, 2020.

CHAPTER 6
It's Not Your Fault

The more you know about climate change, the more you understand it's going to take astronomical efforts and commitment to set things straight. With that growing awareness, anxiety starts to dwell. You start to feel like change isn't happening fast enough, and it's easy to get frustrated with "idiots" who don't get it.

Thank you for saying no to that plastic straw (applause), but what about the plastic cup you're holding that you will happily drive back home with in your CO_2-spewing car, huh? I see you….

Been there, done that—shaming activist busted.

But I've been shamed too, and it's not fun. It took me a minute to understand this, but shaming doesn't help, especially since no one person can change everything that needs changing in the food, oil, automotive, air travel, agriculture industries, etc.

You have probably heard it's your responsibility to care about future generations and save us from climate disasters. You have to acknowledge our missteps and clean up our mess so we can hand over a livable planet to our children. Maybe you've felt this inkling that your existence on this planet is the source of all evil. You see that to be human today means leaving a trail of destruction for which economically challenged cultures and other species have to pay incredibly high prices.

You're not wrong, but it's also not completely right. I will probably be challenged by many people for saying it, but I will anyway. *It is not your fault*. Nor is it my fault that we're facing a climate disaster.

To set a few things straight: Yes, humans are responsible for climate change. (Recent IPCC reports show no doubt.) Yes, a lot of species are facing extinction because how we live leads to the deterioration of natural ecosystems. Yes, poorer countries suffer because richer ones exploit their lack of power. And yes, the future we're looking at right now is quite scary, and we (humans) are to blame.

However, it is not *your* fault.

It is not your fault a million species might leave the surface of the Earth forever, that we're racing toward a two-degree Celsius temperature rise (if not much more), or that wildfires, flooding, drought, and severe weather events continue to intensify around the globe. It's not your fault that we've depleted our soils and filled them with toxins, that families have to flee their homes, or that the air we breathe is so unhealthy that air pollution alone kills approximately 10 million people every year.[2]

It's. Not. Your. Fault.

It's the world we live in.

To be alive today means riding with the devil on a train moving so fast it's almost impossible to get off. It means no matter what we do, no matter how much we care, climate change worsens with our every footstep.

Do you have to drive to work? Do you sometimes travel to see family or distant places? Do you frequent your local grocery store and buy foods imported from other parts of the world? Do you turn the heat or AC on so you can sleep better at night or so your kid can focus on their homework?

Even if you said no to all these things, even if you're some sort of eco saint, do you not own a phone? A computer? A TV? Some sort of technology that keeps you connected with the rest of the world? And again, if your answer is no, how did you get your hands on this book? Unless a friend walked over to your house and dropped it on your doorstep, you probably had it delivered, or walked to a store and purchased it yourself. Either way, it came with a footprint.

The store needs electricity to stay open and customers, who most likely drive there, to buy their books. The books were delivered from elsewhere in the first place, printed with ink on paper that even if it's made with recycled paper and toxic-free ink come with some kind of footprint.

I suppose unicorns exist out there who have absolutely no footprint, people who live in remote places with no contact with the outside world. Many Indigenous tribes still live this way, and I'm sure it's a beautiful

---

[2] Wallace-Wells, David. "Air Pollution Kills 10 Million People a Year. Why Do We Accept That as Normal?" *The New York Times*. July 8, 2022. https://www.nytimes.com/2022/07/08/opinion/environment/air-pollution-deaths-climate-change.html

way of life. But they are probably not reading this book. (If you are, I'm honored.)

In other words, it is impossible to live in the modern world without leaving a footprint. We live in societies designed to fuel climate change. Everything we do contributes to warming our climate. We could blame previous generations for creating this world, but it's actually not their fault either. Not really. Or at least, not intentionally. They just wanted to make life better for their communities, make families happier, enable people to cure disease and live longer, and to simplify our lives so we would have more freedom.

New inventions were made to keep food fresh longer, make lives more glamorous and efficient, and our world smaller and more connected. Everything we take for granted today came from past generations' desire to make our lives better. It's not their fault many of these improvements came with unforeseen, damaging side effects detrimental to people and the planet. Just like they didn't know smoking was bad at first, neither did those before us understand the implications of fossil fuels, artificial fertilizers, or plastics. If they did, their hopes and dreams of a better world got the better of them.

If reports of climate destruction and rising temperatures were denied and in some cases even buried, greed got the better of those who decided to hide them from the world. It wasn't their fault either. It was the world they grew up in, which was designed to fuel capitalism, power, and endless growth.

At the core of our being, we're good souls. We are all made from the same purely divine substance, the same dust that makes up all life, love, and spirit. No one is born thinking, *Let's go; time to kill some ecosystems!*

The problem lies in the world humans created. The world we built over the course of millennia sometimes, unfortunately, gets the better of all of us. Some of us "climate folk" like to say that oil executives are evil, but they are really just people. They have lost touch with the deeper connection to life and are so trapped in the system of money, greed, and growth that their choices undermine others' lives and threaten the planet. It might seem difficult to understand, but it's actually not their fault either.

I know you might be triggered by this. If it's not your fault for making bad environmental choices, and it's not the people in power's fault for causing and then ignoring the problem in the first place, then who is to blame?

No one is, and that's my point. When we stop looking for someone to blame, we can turn a new page. On that new page, we find space to write a new story.

You are not responsible for saving the world because the world we live in was not built to last. It was built on shaky ground with even shakier values, and at some point, it is destined to fall. Like a tower built on sand, a time will come when it can't withstand the winds anymore, and it will start to crumble.

You live in this world, you are a part of this world, you have been shaped and influenced by this world, and you have learned to love this world. Even if you want things to be different, it's often difficult to fulfill those intentions simply because the society we live in does not support it. You want to be more mindful, so you start recycling, only to learn recycling in many cases is a myth. You want to phase out fossil fuels only to learn there's no public transit in place. Even if your intentions are good, the system does not exist to support it.

Yet.

Mega emphasis on *yet*. Because the sooner we start to recognize it's not our fault, the sooner we stop shaming and blaming one another, and the sooner we can use that energy to create a new world. Let's not fight the world we live in now, but let a new and even more beautiful one emerge.

Just because a better world is possible and it's not our fault the world looks like it does today does not mean we don't need change. We need a lot of change, and we need it fast, and you have the opportunity to spark it.

Getting excited yet?

Living from this place of forgiveness for ourselves and others can be quite challenging. We are used to responding to those who are hostile to the environment with the same kind of medicine. Fight fire with fire, so to speak. We meet hostile ignorance and greed with loathing and blame.

It's normal to be angry with the world. (We will discuss how to work with our anger later.) But it's also important to remember that getting angry and acting like those you oppose doesn't work. Only finding common ground will help us move forward.

Imagine waking up every morning for the rest of your life and telling yourself it's your responsibility to right what's wrong and save the Earth from a climate disaster. Do you think that might get pretty tiresome?

My guess is you'll be burnt out within the first couple of weeks. I say that because I've been there. I used to think it was up to me to save the world, and I didn't deserve to be happy until I had done my part. Very little came out of that. Instead, I found myself in frequent clashes and with a mother who kept saying (with increasing pain in her voice), "Honey, you can't carry the world on your shoulders."

Turns out she was right. I couldn't carry the world on my shoulders, but I could carry it in my heart. Isn't that beautiful? The thing about your heart is the more you add to it, the more it grows. Your shoulders will get tired, but your heart will not. It will only grow stronger, more resilient, and more capable of holding space for new change, possibilities, and love.

If we dump the guilt game and instead begin to recognize what an incredibly exciting time it is to be alive, we open the gateway to our heart and its power. If we wake up every morning and remind ourselves we have the opportunity to make history and participate in the cocreation of a better world, we unlock potential we can barely dream of.

It is not your responsibility to "fix" this world. It's also not your duty to find the villain who caused all this pain. If you think it is your responsibility, your heart will soon grow resentful and exhausted. That doesn't mean we should turn a blind eye to all the problems and ignore our fears.

We should speak loudly and clearly about the world we want and the changes we desire. We should share those visions with others, aim for big, bold moves, and hold companies and people in power accountable. We should talk of hopes for a better tomorrow with excitement and proudly share the seemingly small but oh-so-important actions we're taking now to bring that world closer. We should inspire with our actions, speak with conviction, and ask questions so new answers can be found.

But, and here's the caveat, we can do so from a place of curiosity and courage, not from guilt, shame, or fear. We can be the water that calms the fire, and as the hot stones sizzle in the quiet night, wait patiently for new life to emerge from the ashes.

CHAPTER 7
Do I Need to Raise Chickens?

Climate optimism, in my world, is about shifting the narrative on climate change so we can begin acting from a place of courage and excitement, not fear.

This shift has nothing to do with wishful thinking. It's not about denial (although both wishing and denial do play a vital role in our lives as healthy activists—more on that later), nor is it about praying someone with greater power will step up to the plate and do the right thing. You don't even have to be sure we will figure this out. You can still worry and be a Climate Optimist. (I do all the time.) That is what's so beautiful about this journey. You don't have to fool anyone, and the last person you should try to fool is yourself. Be worried, be angry, and allow yourself to feel anxious at times; just don't let it stop you from taking action.

"Action? But I'm not an activist," you may say.

Don't worry; you don't have to join Greenpeace or tie yourself to a pole in the town square to be an activist. I also won't try to convince you to move to a small farm outside the city and raise your own chickens. I won't try to sell you on the dream of giving up modern luxuries and living off the land because no matter how many pictures of cute chickens and idyllic sunrises over dewy fields I include in this book, most likely you are still not going to do it.

I, too, got lured into the idea of an idyllic homestead life. One time it got so far that my husband Arthur and I were looking at buying an inn in the Massachusetts countryside. We had escaped New York during COVID (like so many others), and when we were unsuccessful in buying a house in the crazy real estate market, we thought maybe an inn would be cool. Demand for country inns wasn't as high so properties like that were still somewhat reasonably priced.

We thought it would be a perfect project to renovate the old farmhouse and turn it into a sustainable inn—an oasis for stressed-out city dwellers to escape to, somewhere for New Yorkers to come and get a slice of nature and peace of mind.

Over the course of a weekend, we had it all figured out. I would serve homemade granola for breakfast and he would be the cool and friendly guy at the bar. In the fall, we would have an afternoon soup-and-wine bar with open fires for hikers, and throughout the year, we would host weekend retreats featuring different experiences. In collaboration with local farms, we would host volunteer trips where visitors to the inn could get a feel for what life on a farm is truly like, with everything from feeding chickens to loading hay and milking goats. We spent hours dreaming about our new lives as owners of a country inn. I even thought of Swedish proverbs I wanted to hang around the hallways. It was a fantastic idea.

Luckily, we didn't do it. We realized we were in no shape to run an inn and decided to buy an apartment in New York City instead. Now we're those city dwellers dreaming of escapes to lush countrysides, so if you feel called to open an inn like this, please let me know. We would *love* to come visit.

The point is, I'm not trying to sell you on the idea of a wholesome homestead life. If that's your calling, wonderful. I'm jealous. But if you prefer to be able to hit up the corner store for last-minute coffee, stay where you are. The truth is cities are a mega-opportunity to create meaningful change in fighting climate change. Since cities are so densely populated and account for 70 percent of the world's greenhouse emissions, a lot can be done there. Not only can we tackle the biggest areas of greenhouse gas emissions by addressing cities, but the density in itself creates opportunities for smart solutions and energy-efficient living.

Half of the world's population lives in cities today, and that number is expected to reach over two-thirds by 2050, so urbanization is here to stay. Besides, imagine if all these people suddenly moved to the country—how crowded the "remote" countryside would be. On top of that, all these people would suddenly have to drive long distances to get places instead of the convenient public transit system in cities, so that's not quite the solution for climate change either. It's not about ditching our current lives and quickly picking up new ones. It's about stirring the pot inside the lives we already live to see where we could possibly add some new spices.

How can you become a climate hero in the life you're already living? How can you make seemingly small but incredibly meaningful changes in your home, community, and town to contribute to the shift toward a smarter and healthier world? That is the question I want you to keep pondering and the mission I hope you'll soon take on with passion and excitement. When it comes to creating a better world, we all have a part to play.

But okay, you have a life. I get it.

Before you start feeling too overwhelmed about changing the world, let's backtrack a few steps. I understand you have obligations, responsibilities, personal goals, and routines. You probably have a lot of people relying on you—kids, parents, partners, coworkers, bosses, *yourself*—and it's hard to see how "saving the world" could fit into that schedule. Therefore, I'm not here to tell you it's your responsibility to save the world. As you already know, I don't believe in that agenda.

But being a climate optimist is not about dumping a heavy load of responsibility on your already crowded plate. If anything, it's about creating more space for new thoughts and visions to unfold and making room for something even better. And all you have to do is be a tiny bit more open to change. Take inventory of your existing life and ask yourself how you could do things differently. How can you do things slightly better? From there, magic comes into play, because as soon as you say yes to change, you board a train with unfolding possibilities, and you frankly have no idea where it will take you.

Wouldn't you want to give your life that kind of meaning? Wouldn't you want to look back at these times and say, "Wow, look at what we accomplished. Look at all that change"?

The change I'm referring to starts right here, right now, with a deep awareness of your current reality. You allow yourself to remove all filters and see there is nothing to lose, only to win, and that from here we can only do better. You also understand no superhero is coming to save us. It's up to us to do what we can to change the script of what tomorrow brings.

In that space of deep awareness, a lot of dreaming takes place. You allow yourself to dream of something better, and since you've realized we have nothing to lose, you will find the courage to explore where those dreams might take us. Your curiosity is sparked, and you begin to have thoughts, even if microscopically small, echoing a silent but oh-so-enticing, "What if…?"

With an increasingly attentive ear, you will begin to listen to those "what ifs." Soon enough, you will find yourself starting to believe in slightly better worlds. Small voices inside you will whisper tales of new realities, and from there, excitement will grow.

Curiosity leads to small actions, small actions form new habits, and new habits reshape who you are. That new you will look at yourself and the world differently, and what you believe to be true, right, and possible will begin to shift. As you shift, the world shifts with you, and one morning you will wake up and see it all so clearly.

You are the change.

That is what climate optimism is about. It's about being the change and, in that act, believing in a better world. It is time we shift the narrative on climate change so we can see it for what it truly is—a time with a shitload of information telling us we are not living life the right way. We get to assess that information with our brilliant brains and use that brilliance to redesign our lives moving forward—not for worse, but for better.

We are not here to sacrifice anything. We are here to seek even more abundant lives, better and more beautiful than anything we've seen before.

CHAPTER 8
We Are All One

Climate change understands something we seem to have forgotten—we're all one—one Earth, one people, one magical life force living through space and time.

We may live in different countries, speak different languages, have different skin colors, and follow different traditions, but we're all one. Some of us have summer while others have winter. Some of us grow up surfing while others ski. We have different backgrounds, different expectations, and different dreams, but we're still one.

It can be hard to embrace this truth because so much in our world wants to tell us how we're different. Beyond what's obvious from birth (race, skin color, body type, etc.), we have created societal differences to ensure we stay different. Job titles, zip codes, material attachments, and social clubs help categorize us into different boxes. We nurture a feeling of "us" and "them."

But climate change doesn't care what your zip code is or if you wear Gucci or H&M to work. Climate change doesn't know about our human-made borders. It doesn't know about who did what and who ought to pay and who should be left out because they've done nothing wrong. Climate change is ruthless, and, I hate to say it, more so to the people who have done the least to deserve it, disproportionally affecting poor communities and countries. The people who have done the least to fuel climate change are currently the ones paying the highest price, having to abandon their homes due to flooding or drought or suffer early death from air pollution and other toxins.

Climate change is also cruel to non-humans who never wanted anything. Other than to live in harmony with their ecosystem. Instead, they're starving or are forced to migrate long distances in search of food, oxygen, water, etc. Birds who are known to mate for life are starting to "remarry" because their partners don't return from sea since their fishing trips taking longer due to changing waters.[3]

This is climate change in action today. Not in some dystopian future, but right now. It's hard to see and understand the pain already taking place when we think of ourselves as separate from the "other." Without the deeply embodied understanding that we're all one, the plight of the other may appear as a tragic circumstance we feel deeply about, but we don't have enough personal attachment to do something about it.

If we could edge ourselves closer to understanding "one," we could arm ourselves with an incredibly powerful tool. Out of that oneness, out of that love, we could move mountains, create change, and make the impossible possible. That unity won't just fix the problem in the moment, but set us on a journey of monumental, sustainable change.

Start practicing embodying this truth. Add it as a mantra to your morning routine or save it as text on the back of your phone. Say it out loud to yourself often, and pay attention to whether you're starting to change the way you look at yourself and the world.

We are all one.

Being one with the world shouldn't give you a sense of guilt and/or responsibility. Instead, you might feel as if someone wrapped you in a big, warm hug. To recognize we're all one is to allow yourself to be one with that bigger force again and understand you're never alone, not acting in a vacuum. You understand that we're all in this together, whether we can see it or not.

Nature does not exist outside of you but inside. You're not separate from nature; you are nature.

We may have forgotten, but when we remove all the barriers—mental, geographical, or cultural—we are all one.

---

[3] https://www.bbc.com/future/article/20170808-climate-change-is-disrupting-the-birds-and-the-bees

CHAPTER 9
An Intersectional Matter

Is climate justice and environmental justice the same thing? What is climate justice? How can one provide justice to climate? As you might understand, it's not about giving climate a stool in the courthouse, but about working for stabilizing the climate in a way that brings justice to all. This is incredibly important to recognize because if we don't, we might end climate change, but harm and continue to exploit people, animals, and communities in the process. Therefore, it's not just about finding the fastest, cheapest climate solution, but recognizing that short-sighted, poor decisions brought us to where we are now and we must change our thinking moving forward. As the brilliant Albert Einstein said, "We cannot solve our problems with the same thinking we used when we created them."

That means we have to consider how our move forward can benefit and bring justice to all, the natural world included. That is where environmental justice comes in. For example, if setting up a solar farm means cutting down trees and killing the ecosystem in that area, it's short-term thinking, not long-term, and doesn't include justice for everyone involved. Although they are similar in nature, climate justice and environmental justice mean slightly different things, and it can be beneficial to understand the two and how they're linked.

Climate justice and climate work focus on the fact that our globe is warming due to too many greenhouse gases in the atmosphere. It aims to halt this warming as fast as possible and even find ways to reverse global warming. Climate action means all the things we can do to halt climate change and bring CO_2 parts per million (ppm) levels closer to normal, and in doing so, hopefully, prevent the worst disasters.

The justice part is making sure the underrepresented and often most affected countries and communities are seen and heard. It means recognizing there can only be justice if there's justice for all, and since these people are already hurting, we need to act on their behalf right now. Climate justice also focuses on diversity in voices and recognizes much can be learned from Indigenous people's customs and wisdom, especially when it comes to tending the land. It also means listening to those who live in vulnerable areas who best understand their land's ecosystems and the changes they've seen over the years. When deciding to mine for natural resources, for example, we must listen to and respect the people who speak the language of those lands. We listen to them not just because we respect them, but because they have important knowledge that can help us understand the complexities of their ecosystems and how one ecosystem plays a vital part in our climate at large. The earth has a natural system to balance the carbon cycle, but if we continue to turn natural areas into humanized land, we disrupt this natural carbon system and automatically fuel climate change.

This brings us to environmental justice and how the two are linked. Environmental justice focuses on changes in the environment more than changes in the climate. Of course, changes in our climate inevitably lead to changes in the environment, which is why working for justice in both is equally important and has to happen at the same time. People working for environmental justice tend to focus on biodiversity in an area, preventing deforestation, laws against pollution, or other important work to keep our natural lands safe. It's about protecting the environment around us by recognizing the natural world has a right to speak too and that taking care of nature plays a vital role in our own wellbeing and survival.

Since we're all one, we are also part of the bigger ecosystem, and we need biodiversity to continue our existence. Therefore, we are linked to the planet, just as the planet is linked to us. How we effect the climate and hence the environment around us has a direct effect on us in return. It's all interdependent, and the sooner we see that, the sooner we can begin to make things right.

Terms like climate justice and environmental justice are often used interchangeably, and more often than not, speak to the same thing. Climate justice is referred to as the same work as environmental justice, and in many senses, it is the same work. If nothing else, working for one will definitely benefit the other, so how you use these terms doesn't really matter. Don't get too caught up in the terminology. The important part is knowing it all fits together, and we must focus on all fronts at the same time.

Social justice means climate justice; climate justice means environmental justice; environmental justice means a cleaner, healthier, and better world for all. At the same time, damage in one field is sure to harm other areas of justice as well.

Let's take plastic pollution for example. We know it's bad because we're harming the environment by contributing to plastic waste in our oceans, but it goes deeper than that. By interrupting the ocean's ecosystem and killing marine life, we're disturbing the natural way of life in those seas. Our oceans are incredibly important as an ally in tackling climate change. Serving as the globe's natural cooling system, the oceans absorb the majority of all greenhouse gases, storing fifty times more carbon than the atmosphere and twenty times more than land plants and soil combined.[4] But the oceans need rich marine life to thrive, so the more we pollute and overfish our oceans, the more we limit their ability to cool the Earth.

Similarly, as we continue to burn fossil fuels and warm up our world, the oceans are getting so acidic that fish, coral, and other marine life are dying. This should concern us a great deal since we rely on the oceans for about 50 percent of all the oxygen we breathe.

The warming climate and our dying natural world will continue to degrade each other, but it doesn't stop there. If we take it back to plastics, we should also be aware of what that pollution is doing to people and communities. In places with little to no ability to manage trash and recycling, people have to live with that pollution and can sometimes barely see their river's water for all the plastic, as is the case in the Ganges River in India, one of the world's most polluted waterways. People dependent on these rivers are forced to live in toxic, foul-smelling conditions, intensifying and furthering the effects of poverty.

But plastics don't just create environmental and humanitarian harm once they've been used and disposed of. Plastic manufacturing (which

---

[4] Isaacs-Thomas, Isabella. "When it comes to sucking up carbon emissions, 'the ocean has been forgiving.' That might not last." PBS News. March 25, 2022. https://www.pbs.org/newshour/science/the-ocean-helps-absorb-our-carbon-emissions-we-may-be-pushing-it-too-far

requires oil aka fossil fuels) directly affects surrounding communities. Polluted air leads to asthma and heart diseases—along with the simple fact that people have to live in a toxic environment, which further fuels social injustices and provides another sharp image of how we must change.

By recognizing how interconnected every piece of this problem is, we can take a different look at our daily habits and also understand what a powerful and beautiful part we all ought to play. Because if we can all be part of an interconnected problem, it also means we can be part of a chain reaction of positive change.

Here is where the deep breath of this conversation comes in. When you recognize that if you work for justice in one area, you will automatically fuel justice in all areas, you start to see how it all comes together. You won't be overwhelmed by the many problems we're facing, but instead empowered in your ability to make a difference.

Luckily, there are people passionately working for justice in all areas of climate and environmental justice, and we need all the heroes we can get!

The best thing *you* can do is find what drives you and work for that. And if you don't want to invest yourself in any one area, commit to remaining an advocate for positive change and celebrate and support every win on this exciting journey. When any climate justice happens, recognize how it means justice for us all. When a new law is passed prohibiting further ecocide or pollution, understand it is a win for you too.

As an interconnected whole, we all matter, and we all get to be part of the solution.

CHAPTER 10
Leave a Big Footprint

Everything changed for me when I realized my life's mission isn't about minimizing my negative footprint, but maximizing my positive one. Up until that point, I had thought my very existence was the cause of all evil, and the only way to make it better was to make myself as small as possible. I thought I had to shrink my presence on Earth to one that is barely noticeable. I thought nature could then go on playing its song without my ugly interference.

This view of my relationship to the world paved the way for lots of self-inflicted pain. When you truly believe humans are the cause of all evil, and since you're human, you're evil too, it's pretty hard to wake up in the morning and feel like your life matters.

Instead, I found ways to punish myself so I could feel like I belonged here. I starved myself to feel the pain of suffering animals or denied myself pleasures because I honestly didn't think I deserved them. Who was I to feel happy, healthy, and loved when I knew there was so much pain and suffering in the world?

The first thing that shifted was my understanding of my influence on the environment around me. I began to notice how my energy had a direct effect on others. In my naive, self-pitying existence, I had forgotten I don't live in a silo and that the energy I put out into the world has a ripple effect. But then it hit me—who was I to bring anything but good to a room? Who was I to shed my negative energy for others to absorb?

That was my first peek into the meaning of leaving a big footprint. Because if I can add something beautiful, why wouldn't I make it as big as possible?

Suddenly, I saw my contribution to the human world in a whole new light. But how about the more-than-human Universe? How about the plants, birds, and furry friends I love so much? That one was trickier, but since my heart had already cracked open with some glimpses of an understanding of universal love, the next lesson was easier—and would change my life forever.

It began in 2018 when I developed a deep fascination for soil. Having just read *Kiss the Ground* by Josh Tickell, I brought up soil at every party. And I mean *every* party. At one point, Arthur looked at me before stepping out of the car and said, "Honey, maybe don't bring up soil tonight."

But how could I not? The universe of life that exists right beneath our every step is so fascinating it should be the highlight of any party. I couldn't believe the magic web that exists right below us. A web that we have (sadly) been destroying since the first time we dug a shovel into the ground. But it wasn't learning that in some parts of the world, we have merely seventy years of growable soil left or that agriculture as an industry releases loads of carbon dioxide into the atmosphere that hooked me.[5] It was that we—humans—could do something about it.

Not only can we prevent more damage by minimizing our footprint, but if we step in and make our footprint larger, we can even begin to reverse soil deterioration.

Get it? Not just decrease damage, but increase good soil! If we start working with nature again and learn how to give and take in an equal balance, humans can be the source of life. We can be one with the ecosystem, just as nature intended.

The truth is we have been fed a lie. We were told we are to blame for the oceans being polluted with plastic and for our air being polluted with greenhouse gases, and that we should seriously consider our habits so we can minimize our footprints. Shame on us.

No, not shame on us. Again, it's not our fault we live in a world filled to the brim with plastic without a proper system to dispose of it. It's not our fault that society is structured so it's close to impossible to get anywhere without releasing some CO_2 into the air.

Of course, we know we do all this, and so we walk around with this constant guilt that we should do better, should try a little harder to make

---

[5] Ritchie, Hanna. "Do We Only Have 60 Harvests Left?" January 14, 2021. https://ourworldindata.org/soil-lifespans

our presence here less damaging.

But what if your presence shouldn't be felt less, but more? What if the answer isn't to retreat into shame but to step out into your power? Maybe your existence here on Earth is so meaningful you owe it to the world to leave as big a footprint as you possibly can?

I know it is. I know it because you're alive, and to be alive means to have the opportunity to do the extraordinary. It means you have the ability to make history and leave the world even more beautiful than you found it. To be a climate optimist means to insert yourself back into the cycle of life and ask, "How can I matter? How can I leave a big and beautiful footprint on this big and beautiful Earth?"

Mother Nature is waiting for us to join her squad again, and forgiving as she is, she will welcome us with open arms. We are her children too, just as much as the bees and the trees and the sea turtles. We are nature. We are beautiful and wondrous and the most powerful mammal on Earth. Why wouldn't we use that power to do as much good as we possibly can?

The time has come to step out of our shame caves and into empowered action. Let's leave a big footprint, and let's do it now!

CHAPTER 11
Capitalism—Good or Bad?

In Western culture, since the inception of trade, the idea that something can be exchanged for the value of something else has been hated and loved. Those who had enough loved it; those who did not hated it. Today, we live in a similarly skewed relationship to capitalism and money. On one end, it seems as if capitalism has taken on a life of its own and turned into a monster that cannot be stopped. Capitalism exploits people, ruins natural ecosystems for capital gain, and lures people into buying traps they can't seem to get out of. The idea that capitalism is the source of all evil is no news to many of us, especially those who have long been immersed in climate justice work, and it's not rare to hear that we should ditch capitalism altogether.

I think that's a dangerously simplistic way of looking at an incredibly complex problem. Yes, the idea of being free of capitalism altogether may be attractive, but how realistic is it, really? When people say, "We must end capitalism," I can't help but look at the clothes they wear and the iPhones in their hands, and then think about all the ways they have benefited from the capitalist system that day. I understand why ending capitalism sounds like the only solution in many ways, especially since it is the bottomless hunger for more, more, more that has us kept in this constant downward spiral of ecological disaster, but we're missing the fact that the world today is built on capitalism. We can eat, dress ourselves, heat our homes, and see a doctor when we need to because of capitalism. Without it, we would need to go back to the old days when everyone made their own clothes, grew/raised their own food, and found fuel for a fire to warm themselves.

My guess is few of us are willing to go back to that, and even if we did, our infrastructure simply wouldn't allow it. Imagine people living in apartments trying to start fires in their living rooms? How many buildings would burn down? Who would come put those fires out? Without capitalism, it is hard to make money, and even harder to support societal services like firefighters and their expensive equipment.

We have created a pretty great reality for ourselves. Let's begin by recognizing that because we shouldn't take it for granted.

Dismantling capitalism altogether would mean throwing ourselves into possible disaster. Therefore, let's skip the idea of ditching capitalism and instead look at how we can shift the narrative on how we think about money, value, and products.

As with anything powerful, if it can do lots of harm, it can also do lots of good, and the same goes for our economy. That means if we can change our focus from how we can do, make, and have as much as possible for as little as possible, we can enter a capitalist system that recognizes and supports the fact that there is enough for everyone. It means we can shift our values from constant growth to circular growth and find ways to keep products and materials circulating in the system instead of ending up in a landfill. A forest is always growing, and there's constantly new life all around, but that doesn't mean the trees continue to get taller. Nature understands circular growth, and we can too; all we have to do is shift our mindset and aim for a different kind of better, a different kind of more.

Shareholder vs. Stakeholder

Luckily, this shift is already on its way. More and more companies are looking at how to embody stakeholder values instead of shareholder values, and in doing so, reframe economies for the better. The old way of growing a business was to strive to make shareholders happy—in other words, to make the rich people who invested money in the company richer. Therefore, being able to show constant growth and yearly and quarterly returns was the heart of any successful public business. Not only is it a single-minded pursuit, but it also stifles a lot of good initiatives. If a fashion brand wants to switch to more sustainable practices, for example, but must make a significant investment to do so, it may not look too great on a profit and loss report. Shareholders may balk at the change, even when calculations prove that making the change will bring down costs in the long run while also helping the environment. It doesn't matter if it means bringing quarterly earnings down. That's the very definition of short-sighted. This, quite frankly, is the major drawback of capitalism and the reason the world is continuing to head toward its demise.

Fortunately, today many companies are starting to think about more than investors' profits and give weight to every stakeholder's needs. A stakeholder is anyone directly or indirectly affected by the company's business. It can be an employee working at the company, that employee's family and future, and the community in which the company operates. Stakeholders include customers, the areas in which those customers live, and the ecosystem that surrounds a business' activities. Focusing on stakeholder value, you don't just look at how to make money, but how to operate in the best interest of people, animals, and nature. It means making needed sustainability investments because it's necessary for the survival of the business and, of course, our planet. It means making conscious decisions about how investments will affect others, which is why many are referring to this move as the Conscious Capitalism Movement.

Looking at businesses from this lens may serve as a beautiful opportunity to create substantial change. Since we are so dependent on capitalism and the systems we've created, we can turn this massive force toward creating something good. Businesses have the power to change communities, improve lives, and invest in environmental projects that can restore nature and give it a chance for survival. Businesses can also fund new solutions and explore better materials, products, and processes that allow things to circulate or return to the earth in non-harmful ways.

If we recognize capitalism is fueled by our human desire for something better, we get to decide what that better is and what it looks like. We get to say we value long-term, sustainable business practices over short-term economic gains because, quite frankly, how we make the money actually matters.

I understand that, right now, capitalism is like a monster that doesn't care about the planet or people, but that doesn't mean it can't change. Give up on capitalism altogether and the people who don't care for equal, sustainable values will continue to reign. If you instead say, "Hey, I believe in this system, and I think it can be the source of really great things," you will have a voice in the game.

The system is shifting, and it will continue to shift for quite some time. Already incredible companies are providing products and services that are good for people, profits, and the planet. They are fueled by innovative and passionate people who want to see change and have gone to great lengths to achieve it. We owe it to these people to honor their courage, so please, whenever you can, support their efforts.

At the same time, some bigger whales are starting to turn as well and are also recognizing the need to shift their values. They might take longer to turn because their existing infrastructure makes it harder to do so, but at the same time, because of their market share, they have a lot of clout when it comes to making substantial change. A big company shifting 20 percent of its products to sustainable practices can make a significantly bigger change to the industry at large than a small brand producing 100 percent of its products sustainably. And when the big company sees people valuing their efforts, they will look at that data and continue to do more.

Who should we support? I'd say both. Give your powerful money to the small business owners who are paving the way to a better world, but also celebrate the efforts of the giants deciding to change their ways. We won't reach a circular and sustainable world tomorrow. A shift needs to take place, and we get to help fuel that shift—spiritually and economically—with every purchase we make.

Don't give up on capitalism. If we do, we'd lose a huge ally on our journey and a big opportunity to speed up our transition to a better world.

CHAPTER 12
Tone Down the Urgency

In the early spring of 1991, a little girl was born with more willpower than a horse on hormones. Yes, that girl was me. I came into this world with a fire engine roaring inside me, and I knew what I wanted from day one. A typical Aries, I am simply not one to sit around waiting for things to happen. I take matters into my own hands to make sure they do!

That's why what I'm about to tell you is just as much a reminder to me as it is to you. Although a lot of what I share in this book has somehow been channeled from inside me (many times while gazing at the horizon from a mountaintop I just summited), this message came from without, almost as a plea for me to change my ways. Had I received it in person, instead of over the phone from a friend, I'm sure it would have come with two sturdy hands on my shoulders and two eyes piercing into mine. That message was:

Slow down.

I'm not good at slow. I like things at speed. I like to see things moving! Being the "go, go, go!" person I am, this lesson has been the hardest one to learn. How can I slow down when I know how much the world needs my actions? How can I "chill" when I know climate change is rapidly speeding us toward our demise?

The anxiety that kicks in every time I try to give myself a break is difficult to handle. That is why it's easier to fill my time with stuff. Even sharing a post on social media to educate others about the problems with plastic makes me feel better because at least I'm doing something. But what am I actually doing? And is the planet better off with yet another post circling the internet for a day or two before shortly getting buried in the feed?

Truth be told, I crave slow, and I think the world craves it too. We are dying to breathe more deeply, to see the world more clearly, to better understand ourselves and the deeper wisdom whispering from within. The world doesn't need faster. It needs more space, and that space can only surface when we allow ourselves to slow down.

When it comes to a sustainable future, we need it more than we can know, but it's time we start toning down the urgency on climate change.

"Wait!" you may be thinking. "What did she just say? Tone down the urgency?" I am fully aware this statement makes very little sense. I will not argue we don't need rapid changes. We do need a lot of change, and we absolutely need those changes to happen fast. Preferably, we will have cut all global emissions in half by the end of this decade (2030), and then we're on a new sprint to cut even more by the next. However, to succeed in doing this, we need to tone down the urgency so we can focus and ensure we do it right.

The friend who told me to slow down was Arjuna Ardagh, author, executive coach, and founder of the Radical Brilliance Project. Specifically, he told me:

Since saving the Earth from a climate disaster is of such importance to you, it's even more important that you slow down and act from a place of clarity and intention. I will give you an example. Let's say you're a brain surgeon and someone you love dearly comes into the operating room. You care deeply about this person and want nothing but to keep them alive. Are you with me so far?

Now, will you do all the procedures you always do—clean every piece of equipment thoroughly, prepare your nurses, set the room up for surgery—or will you skip all the necessary steps and rush right in because you're so afraid they won't make it? No, you would do everything properly and according to routine. Why would you treat your work for climate change any differently?

Boom. Message received, and (most of the time these days) fully implemented. I understood what Ardagh was saying. I could see how my constant sense of urgency wasn't doing any good. If anything, it was keeping me in tunnel-vision mode where I was unable to carefully assess the situation.

We are in such a critical situation right now that we don't have time to rush into the next thing. We must begin by slowing down so we can move with intention and clarity toward a world that is not just different, but better. If we don't slow down first, how will we know if our actions will lead to better outcomes or continue to fuel new problems down the line?

To solve the climate crisis, we need to find a deeper connection to our hearts. Charles Eisenstein explains this beautifully in his book *Climate: A New Story*:

Rational reasons are not enough; the ecological crisis is asking for a revolution of love. If there is another way, then the habit of fighting becomes an obstacle to victory. In the case of ecocide, the mentality of war is not only an obstacle to healing, it is an intimate part of the problem.

For many of us, fighting climate change seems like a war, a battle we must continue to fight. We get frustrated when elected officials don't do enough, and we happily shed that frustration on the people around us. Tired, hopeless, and sometimes even a bit resentful, we feel as if there's no way in hell we will ever get this right.

That is the fighter in you talking. Beneath that fighter lies a powerful soul powered by a strong heart that can't wait for the chance to say a few words. That heart knows of wisdom so strong and so powerful it could change the world.

We don't need more fighters in our world today. We need wise hearts moving forward with one goal in mind—a beautiful world for all to thrive in. People aren't selfish in their thoughts and actions because they don't care about other people; they are selfish because they carry fear in their hearts—fear they won't have enough, fear of rejection, failure, and pain.

With a wise and powerful heart, there isn't room for fear, only love. It may sound cheesy, but the simplest and most profound things usually are—it's time for a revolution of love. Let's tone down the urgency and get to grounded, guided, and intentional action.

CHAPTER 13
You Are a Tiny Fish in a Huge Ocean

Sometimes you feel incredibly small. Why? Because you are small, very, very small. Think about it—you are *one* in almost 8 billion people. It's not like the odds are in your favor, and as if that were not enough, the power play between those 8 billion people is far from equal. A few people have lots of money and a lot of power over big corporations that have even more power. Let's call those people the big fish—the sharks. The sharks swim the oceans thinking they can go after whatever they want without repercussions.

That is most likely not you. You are a tiny fish in a huge ocean and you will often feel incredibly powerless and question if what you do matters at all.

Why should you bother bringing your own bag to the store when the person behind you will gladly use ten plastic bags—wait, double-bag it, "Just to be safe," and make that eleven? Why should you carry the extra weight of a reusable mug in your bag when everyone else in the coffee shop mindlessly reaches for the disposables?

Sticking to your zero-waste habits can be exhausting, and if you've already adopted some or many of them, you know exactly what I'm talking about. (I see you nodding.) Or if this is all new to you, you will understand it all way too soon.

The question remains—does it matter? Do you, as the tiny fish you are, matter in this ocean of consumerism, politics, and corporate agendas? The simple answer is yes, and don't you for a second believe you don't. I wouldn't be writing this book if I didn't believe in the individual's power to make a difference, and I have my own life as an example to prove it.

I'm not boasting. I'm actually awed and mesmerized by the fruits of my labors that I've harvested over the years. I thought I was working in a silo when the truth was everything I was doing had either a direct or indirect effect on the world around me. Old high school friends whom I haven't spoken to in years have reached out to say I have inspired them to make sustainable changes and to thank me for their improved mindset and lifestyle.

Arthur, who's the most stubborn person I know (a true bull), will now come back to grab the thermos he forgot for his coffee. When we met, he was completely oblivious to the effect of throwaway cups and plastics, ordering two or three Starbucks coffees per day.

You have that effect too. You can plant seeds, shift minds, and expand hearts just as well as anyone. That doesn't make you powerless. That makes you a force to be reckoned with. Yes, you might be a tiny fish in a huge ocean, but don't forget there are other fish out there. And if we come together in schools and start swimming in the same direction, we're not just many; we're creating completely new currents to follow.

The trick is to avoid trying to change *the* world. That is the kind of ego-tripping only the teenage version of myself was up to. I've learned I can't change *the* world, but I can change *my* world, and that in itself is much more powerful than anything I can imagine.

Don't try to change *the* world. You'll just fail and feel like the sharks are out to get you. Instead, give your everything to changing *your* world and watch in awe as *others'* worlds start to change as well.

CHAPTER 14
Fueling the Transition

There will come a day when you barely remember the smell of cars whipping by you. It will be a beautiful day; the air will be much safer to breathe, and you'll be able to hear the birds and bugs instead of all the industrial noise. There will come a day when you don't spend time looking for parking or waiting in traffic. You'll be able to hop on an ultra-fast electric train with boringly on-time schedules to get you where you need to go. There will also come a day when cities are predominantly green instead of gray thanks to all the urban farming and greenery installed to rebalance microclimates and bring food production closer to home. It will be a good day because life in the city will be even more pleasant than it is today, with room for even more culture, music, and all the things we value so much.

That day will come. I know it in my heart. A better world is possible, and the sooner we start to believe it, the sooner we can get there. But what we must remember is the "getting there" part. We have to realize we can't just ditch this world and start living in a new world tomorrow; we must fuel the shift—both economically and spiritually—that will take us there.

That means celebrating companies and individuals taking bold steps in the right direction. It also means continuing to ask questions—to companies, elected officials, neighbors, and ourselves—so we can slowly let go of the reality we know and weave the web of a different tomorrow.

But when you wake up to the realization that this world isn't working and we need to quickly make the jump into a different tomorrow, it's easy to get anxious. Like a child on a road trip, you keep asking your tired parents, "Are we there yet?" You know how urgent it is to act on climate change and understand the steps we need to take to get ourselves out of this mess. You don't have the patience to wait.

However, this is where the biggest challenge of our times comes in and where we get to truly step into the light as heroes. It's not the vision of a better future that's worth celebrating. It's the courageous and tireless work that will nudge us ever closer in the new direction, the never-ending faith and commitment to keep going.

You will often feel frustrated. You will want to give up because, on the surface, it'll seem as if we're not moving forward. More often than not, you'll actually think we're moving backward, and all the work you're doing is for nothing.

But when you start to really pay attention to the shift, you'll see we're not going backward at all. In fact, you'll realize that, holy smokes, things are definitely moving forward. You'll see wonders in small, almost unnoticeable changes. You will see the changes that, if you look back over time, have actually built up to huge shifts in the world you know today. New trends, new lifestyles, and new products in the grocery store are all signs of the shifts we're undergoing.

You'll also recognize that a group of people speaking up and putting up resistance is actually not a backlash, but an important part of dismantling the system we know today. Breaking something apart to build something better gets messy at times. That is what the breaking apart part looks like—like a big freaking mess—and it should never be mistaken for going backward.

We're living through that mess right now. Unfortunately, many are being hurt in this process, and even more hearts will be confused, but that is why it is so important to keep your vision clear and your head strong while remaining a guiding light for others to follow.

When we seem stuck in the dark, we don't need more darkness—we need light. Be that light. Keep fueling the transition. Stand up for what you believe to be right, and pay close attention to the shifts all around you. Use what you see as your fuel to keep moving forward.

And if you don't think the shift is happening fast enough, think about what you can do to fuel it. Maybe you can sign your building up for curbside composting or help your neighborhood get involved with soil restoration and regenerative farming. Perhaps you qualify to sign up for solar panels on your roof and can become a trailblazer in your community, incentivizing others to sign up as well.

It's easy to look to politicians to fuel this shift for us, or for big corporations to step up to their responsibilities and do the right thing. But you hold the biggest power. You are the strongest motor for activating change where you live.

This shift, this beautiful transformation into a new world, will have to happen everywhere at once. It is not a top-down sort of change. It's a revolution that starts with the people and spreads fast. The world will have to catch up.

Are you ready for the task? Are you willing to become a trailblazer, a change maker, a hero of our times?

Beautiful. Now, let's get into the deep inner workings of becoming a reliable light warrior, a revolutionary soldier to be reckoned with.

CHAPTER 15
From Responsibility to Opportunity

I hope by now you've come to recognize the importance of shifting the narrative, not just on climate change, but also on the everyday stories we tell ourselves. We have to move from a place of doom, gloom, and disaster to one of opportunity, curiosity, and hope. We have to stop worrying about what could be if we don't act and start getting excited about what could be if we do. In its simplest form, we have to find the courage—and the curiosity—to choose change.

You do this best by getting rid of the responsibility stamp on your back. If you walk around constantly reminded of what a huge responsibility you have to fix climate change and hand over a livable future to your children, there will come a day (and that might be very soon) when you'll crack. You will feel frustrated, powerless, panicked, and overwhelmed. You will think to yourself that there's no way we can do this, and you will begin to doubt your ability to make a change. The responsibility to save Earth is a heavy one to carry, and I might be naive for saying this, but I don't think anyone who does so for a long time will succeed.

Instead, start acting from a place of childish excitement, joy, and curiosity. Tell yourself you live in a special and meaningful time, and you can make your life special and meaningful too. This means that every single morning you have another opportunity to create change. For to be alive today means being able to participate in the shift toward a better world, and in doing so, make history with everyone around you.

You are not responsible for saving the Earth and humanity. You do, however, have an incredible opportunity to question everything you know, remain in a state of childish excitement, and continue to create positive, meaningful change.

The choice is yours. Do you want to change the world?

I can't tell you what to do, but I don't think you're foolish enough to miss this opportunity. What I share with you in the following chapters should convince you, if not excite you, to get yourself into the driver's seat and head toward a better, healthier, and more sustainable future.

> *"We have the choice to use the gift of our life to make the world a better place—or not to bother."*
> — Jane Goodall

PART TWO
CHOOSING CHANGE

CHAPTER 16
Choosing Change

"All we want in life and all we want for the world lies right behind the other side of courage and conscious choices."

Setting intentions is important. Allowing our minds to wander and our dreams to tell tales of better worlds is also critical. However, the only way to materialize that change is through courage and conscious choices—through *action*.

Taking action doesn't have to be a sprint to the other side. Building courage and mapping out change takes time, and it's a process that's woven by many threads. One choice will not change the world or even your life. But a repetition of those new, conscious choices, will. They build upon each other, like stones in a wall that need to be laid on top of each other to eventually build a house. Once those stones are laid and all the walls are there, you need to build upon that foundation, adding windows, doors, a roof, and paint. And then, when the house is completely finished, you need to fill it with furniture and décor. Once it's furnished, you need to fill it with love. If not, it's simply a house and not a home.

Just like building a home takes patience, commitment, and understanding that stacking one block on another will lead to something substantial, so conscious choices and courage will build the change you wish to see in your life—and in the world. One conscious choice is a great step. It's laying down the first stone. But it's not yet a home; it's not yet a place where you feel comfortable and safe. To get that, you need to keep building.

With each choice, each building block of your journey to change, you will feel more and more comfortable. You will start to see the hint of the shape of something concrete and the feeling of completion will be more real with each day. Soon you can start to envision your life in this new home, imagine how you'll feel waking up in the morning, and dream of who you will become with this new stability because once the project is finished, you'll be a new person. You'll be someone with a home.

Think of your change journey as nothing less. Making bold changes and creating a new identity is one of the most fulfilling feelings there is, but you won't feel at home until the work is done. The first few stones will be exciting—they mean the project has started. Then, after a little while, you will begin to see lots of hard work but still no home. You can feel how things are starting to shift inside you, but you're not yet where you can close the door behind you at night and go soundly to bed. It takes courage, patience, and time to land in your new spiritual body, and that journey can be both challenging and uncomfortable at times. But it's always worth it because, in the end, you will have a home—not just a pretty house, but a home.

This part of the book is all about choosing that change and the mindset and courage that goes into creating your new vision. We will cover psychological reasons for our relationship to change and why we aren't acting more on climate change, or at least why we haven't done so in the past.

You will get to know and understand yourself a little better and then have the chance to ask yourself who you want to be and what choices you want to make moving forward.

To cocreate the better world we all know deep inside is possible will come down to the work covered in the following chapters. It will come down to not just the courage, but also the curiosity to continue to choose change. It's about building that new home and holding the clear vision of one day living there, closing the door, and going to bed soundly.

What you're about to learn can be applied to any positive changes you wish to make, and you should make them—becoming the healthiest and happiest version of yourself can only serve the world for the better. But I hope you will also seriously consider adding your building blocks to the bigger work and recognize this world is not a beautiful home unless it is so for everyone. A new order is needed and the vision for the future is mapped by our dreams and hopes. We all dream of an even better world.

CHAPTER 17
Generational Amnesia

People from nations old and new have contributed to the society we take for granted today, one where a warm car will get you to the store where you can select food packaged in big frozen boxes from lit-up shelves and cook it that same night for dinner. (I'm well aware this is not the reality for all people, but for many places around the world, it is.) It wasn't always like this, and a lot of sacrifice and courage has gone into creating this world—courage and hard work we don't often think about and that are easily forgotten.

When I was in middle school, I had my first introduction to the history of food availability when I wrote a paper on the small village where I grew up, a small suburban town in the countryside of Sweden. Smart as I was, I decided to cut some corners and go directly to the source. I decided to interview the people I knew would have the most to say about that place—my grandma and grandpa. For all I knew, they had lived there forever, and as I interviewed them, they confirmed that belief. They had in fact lived in that same area their whole lives.

This school assignment turned out to be one of my all-time favorites. It was fascinating to hear my grandparents talk about what the town had looked like when they were young, how instead of two competing pizza parlors and a closed post office, there was everything from a bakery, a butcher, a cheese store, and a cute cafe where people met for daily gossip and coffee. Back in their day, people didn't drive to town to go shopping. The shopping was right there.

Then Grandpa went farther back in time and told me about when he was little and his father would take him and his siblings to the next town in a horse-drawn carriage every Sunday. They would travel for hours to stock up on food for the week. Sundays were the only time they had contact with the rest of the world. The rest of the week, it was just them on the farm.

Once a year at Christmas, they killed a pig and had a big bath in a wooden tub outside. Grandpa laughed about how they always fought over who got to go in the water first because it was definitely cleaner before everyone else had a chance to dip in. Life was simpler, for sure, but also tougher. And although there were sparks in his eyes as he told me about his childhood, I knew he highly valued his blue Volvo and the ability to drive to town in the world we live in now.

We have come a long way in a short time. In only two generations, the lives of middle-class people in Sweden have improved a lot. But it's not just Sweden, nor is it purely the Western World. Across the globe, we've brought people out of poverty and into higher standards of living. Only six decades ago, in 1965, the number of children per family and percentages of child death within each country were very different. Back then, you could clearly see a gap between what we refer to as the "developing" and the "developed" world, with fewer children and lower child mortality rates in wealthier parts of the world. However, that map has completely changed, and if you look at the statistics today, there is almost no difference between the "West" and the "rest" when it comes to healthy family sizes and the chances of a child living past the age of five. In merely sixty years, a lot has changed around the world and clearly for the better.[6]

However, this growth has come with its consequences. In only two centuries, we've gone from 280 parts per million (ppm) of CO_2 in the atmosphere to 432.72 ppm, as recorded on May 23, 2022. And the trend continues upward. We reached a level of 412.2 ppm in 2020, despite the economic slowdown due to the COVID-19 pandemic, and we've already gone up another twenty points since then. The world's ecosystems are taking a serious hit from unnatural warming, and one can only imagine where we will be if we continue down the same road. You don't need a lot of imagination to know it won't be pretty.

One would think since we now have access to the entire world in a way my grandpa on the farm could only dream of, we would have a much better understanding of what's going on, spot what isn't working, and fix

---

[6] These stats are shared in Factfulness by Hans Rosling.

what's wrong. With information at our fingertips around the clock and the ability to travel to all corners of the world, shouldn't we be well aware of the damage we're causing?

Of course, we are aware, but we get used to new things quickly and easily forget how things were before. It's nice to live in ignorance of the worlds we've lost, but it's to our detriment in many ways. Most of the time, we don't even know what we're missing. We've simply forgotten.

Even more dangerous is when things happen so fast the next generation doesn't know what the world used to look like a decade or two ago. Generational amnesia is the idea that from one generation to the next, the memories of a lost world simply get forgotten. My generation doesn't know what the streams used to look like when they were full of fish, or the sound of thousands of birds swirling above our heads.

When I try to remember, I think I recall a childhood with a constant buzzing of bees and other pollinators. It's hard to remember, but I'm pretty sure it's quieter on that front these days. Pollinators still exist, but they're not as many, and the eerie silence left behind in their absence sometimes scares me.

When a whole generation doesn't know to react because they don't know the difference, things become complicated fast. Suddenly, we rely on the elderly to pass on tales of how things used to be and infuse enough love into those stories for the youngsters of our time to want to rise up and get to work. In a world with so many distractions, that's a lot to ask from Grandma and Grandpa. And that is if they can even remember how things used to be. Because we actually get used to something new very fast and start forgetting about the old.

Very fast. Very, very fast.

What's funny about humans is we tend to make a lot of fuss about change when, as a matter of fact, we're incredibly good at moving on. In the 1990s, people were convinced the internet was either going to be a fad or it would end the world. Only the people who thought it could become something truly extraordinary were right. A few years ago, imagining a world where whole families would coexist under one roof and work from home may have brought a good laugh to many people, but in post-COVID times, we know that's the reality (perhaps even a preferred and chosen reality) for families of all sizes. We are, once again, in new times, and it happened fast.

The reality is that, as a species, we are really good at getting used to new realities and moving on. I see that as both good and bad. It means we have the capacity to adapt to a changing world and thrive in new circumstances, but it also means we become complacent in fighting for the things that shouldn't change at all.

If you were to put two carrots next to each other, one found at the supermarket today and one pulled from a farmer's field one hundred years ago, and compare the two, they might look fairly similar. Although I expect they would be slightly different colors, especially the vibrance of the orange, they would look pretty much the same. Narrow on the bottom, thicker up top, with green sprouts coming through the crown. The farmer's carrot would probably be covered in dirt, but that's only because someone cleaned the other carrot before it ended up in a pretty pile in your local grocery store.

But don't be fooled. As with most things, you shouldn't judge by appearances. The similar-looking carrots will not come with the same promises of fiber, magnesium, and vitamins. In many ways it will not even be the same food. Why? Because the *soil* they've been grown in is comparably different.

The sad reality is we have depleted the Earth's soil so much that it's almost completely devoid from nutrients. Nutrients are not born in the carrot itself, but just as a fetus gets its food from the mother's body, so fruits and vegetables get their nutrients from the richness of the soil. Due to decades of fertilizers and other "modern" farming practices, we have "killed" the soil. Although modern farming practices have sped up production, it's come at a price. It might look like we're getting more food because the harvests have increased remarkably. But if that carrot has no nutrients, we are, in fact, getting less. The soil that used to be a magical place where life of all sorts was born is looking more like a pile of dirt in which we plant a seed and hope with enough toxic "help" it will grow into something resembling food.

What does this mean? It means we might think we're eating healthy, nutrient-dense foods when in reality, we're eating a shell of something that was once rich and vitalizing. We're not getting the minerals, vitamins, and other important supplements our bodies need to function at optimum capacity. It also means we've gotten used to feeling less good—less energized, less rested, less motivated, less happy, less…less like humans should feel. We have gotten used to being more of a shell of a human than a true human simply because we've created a world where the richness we were once used to is simply gone.

Or at least it's quickly fading and we're not paying much attention….

Although this fact alone is a reason to freak out a little, recognize the power we hold in our ability to get used to new realities and worlds. If we can get used to *less*, we can also get used to *more*. If we can silently sit on the sidelines and let a world of depletion seep through the cracks without paying attention, imagine the world we can create when we aim for *completion* instead.

This section of the book is all about getting comfortable with the idea of choosing again and recognizing that as a species, we're brilliant at pivoting and finding roots in new realities and worlds.

I want you to start paying attention to how the idea of "change" might hold you back, and challenge yourself not to fear it, but to embrace it with all your heart.

A time of reckoning has come when we have to find a third way. We can no longer hold on to the past and simply cherish the memories of older generations, nor can we hopelessly tug along into the lesser world we're on track for now. Instead, we must bravely find a third way, a way of change, but one that brings *more*, not less, and that offers the promise of a better world.

How good can you become at questioning your current reality and taking bold and empowered steps into the unknown? How willing are you to explore how much more this world has to offer, not just for humans, but for all species? Nurture the courage to choose again and practice making that courage part of your daily routine—part of who you are—and you will be the hero future generations will look back on and truly celebrate.

"We will always be grateful for those who had the courage to question everything."
— *A history book from the future*

CHAPTER 18
Two Different Futures

How every morning turns into a full-blown circus, Sophia doesn't understand. Did she really sign up for this when she chose to become a mother?

"Sarah, Aaron, come on. We don't have time for this! Please, help me, Mark," she pleads with her husband. "I need to get them out before they miss the bus. I can't drive them again today; I'll miss my morning meeting."

"Listen to your mother. Get down here right this second!" Mark turns away from the mirror and finishes his tie in an effortless way that only someone who's done it a million times can. "Sorry, sweetheart; that's all I can do. My conference call is about to start."

"It's all right. Thank you...." Sophia gives him a kiss on the cheek before he dives into the soundproof office, a room they invested a lot of money in exactly because of mornings like these. She fills her lungs one more time: "You heard your father! Come down here and get your masks on or it will only be dry food for the rest of the week."

It stings inside when she says that, a pain in her heart that is so hard to accept but even more difficult to ignore. Sophia remembers the days when she could walk into any grocery store and piles of fresh fruits and vegetables greeted her in a rainbow of colors. She never fully appreciated it then. Now she would die to get back to that world. Now it's hard to come by anything soil-grown at all, and if you do, it's ridiculously expensive. Very few can afford it, and although no one wants to really talk about this new reality, it's evident all the packaged food is not good for their health.

People just look more...pale. That's the word, pale. Like a part of their light source is missing. All this indoor, lab-grown food can't be good for people, no matter what they say, and she fears her children's future and the future of their kids. Who knows, maybe by then there won't be any soil left to grow food in at all? And what would that do to them? What would they be as a species when all connection to the natural microbiome is gone?

Sophia doesn't want to go down that road, but more often than not, she can't help herself. The signs of their inability to act in time are ever so present. Her kids have to wear masks to walk to the school bus most days because the air is simply not safe to breathe. They can't just open the windows and let the breeze in on warm summer days because it would take hours for the filtering system to get the air quality back to safe. They check their storm alert system daily because you never know—storms can roll in in a matter of hours. Sometimes they're so severe that lives are at stake.

Sophia feels tears build at the corners of her eyes as she acknowledges the absurdity of it all, and what an unsafe and unsound world they're handing over to their children, Sarah and Aaron, her angels. Her little munchkins have so much energy and life, but they will never be able to experience the magic of nature she had when she was little, never be able to roam the woods or swim the river when they get hot, not unless they wear masks and wetsuits.

The toxins. That is what they live with now. And the fear, the constant hovering fear. What if it's not safe....

"Sarah! Aaron! NOW!"

I'm sorry about that. I know you chose a book on climate optimism, not one on "Doomsday 101." However, as a climate optimist, it's my duty to, on occasion, visit one possible future we could face should we choose not to act. If we don't know what we're trying to *avoid*, being motivated to work toward something *better* is going to be hard at best. I might even go so far as to say it's near impossible.

We're a simple yet complicated species. We are incredibly good at getting used to new things yet very bad at consciously activating that change. We either need a goal, a carrot, or an anti-goal so scary the mere thought of it has us leaving town.

Of course, change is happening between these two extremes. Change is incessant; it never stops, but we don't pay much attention to this flow of life. It just is. It's the kind of change where you look back ten years later and say to yourself, "Huh, look at where life brought me. I never would've thought of *this*."

In that regard, we're expert changers—all of us are. Life is simply a constant flow of change whether we like it or not. However, sometimes life asks more of us. Sometimes we have to step out of the self-driving vehicle and get behind the wheel of the powerful sports car. These are the times when we need change and need it fast or else we'll end up in a world we would not have chosen.

The tricky part with our world today is we need to step into that sports car and start speeding into a different future, one that is very different from the one we're heading for right now. But it can be hard to see it. It's difficult to see because it's ongoing, slow-moving change that is bringing us ever closer to where we don't want to go, and we may not realize our direction until it is too late to change it.

That is why we need to do a little future travel in two different directions. We need to take a jump into the future we *don't* want and get a taste of the world we'll end up in if we don't take control of the wheel (something similar to what I just explained with Sarah and Aaron), and then take another plunge into the future of our dreams. We need to know what we don't want *and* what we do want so we have a clear understanding of the mission at hand. What we don't want is to continue snoozing away in the self-driving car without paying any attention to where we're headed, no matter how nice the ride might seem. Once we end up at the edge of town with nothing but a stinky landfill and an abandoned taco shack, we will wish we would've taken control a long time ago when we still had the chance.

The reason we need to be highly aware of both future outcomes is we need to understand both to trigger change.

Did your parents nag you about doing your homework? Mine did, even if I was the study nerd who overdid it every single time. I think it's in a parent's DNA to bother their kids about homework, like they have a dormant gene that grows stronger as the kids grow older, and that only intensifies with peer pressure from other parents.

If you had a parent or mentor who took responsibility for ensuring you made something of your life, you may have come across the "this versus that" argument.

"Do your homework so you don't end up [fill in the blank]." To avoid offending any important and much-needed professionals out there, I will let you finish that sentence for yourself.

When we study for a test in school, we usually have two incentives inspired by a parent or ourselves. The first incentive is what we *don't* want (the anti-goal), and the second is what we *do* want (the goal.) Both are equally important.

Let's say you don't want to keep flipping burgers forever. Sure, it's great now, but at some point, you want to move on to greater things. Being stuck flipping burgers your entire life is your fear (or your parent's). However, if there's no desire on the other end of all your hard work (studying all those hours to get good grades so you can go to college and earn yourself a degree), it will be hard to find the motivation to keep going. Since hard work is required to move forward, you need more than just an anti-goal to avoid. If not, you might give up and say, "Screw it. Flipping burgers is not that bad. After all, I get free food and cuties stop by all the time."

The anti-goal will slowly start to be less of a deterrent if the effort does not match the reward, which is why you need something attractive on the other side to motivate you to continue. You need a career you want so all the hours of hard work will be worth it in the end. This is where the carrot comes in, and where you need to start envisioning something even better in your future.

But just as your dream presents a carrot to strive for, that dream can also lose its power if there is no stick on the other end. If the work you need to do to reach your dream is (again) both hard and time-consuming, you might reach a point where (again) your current situation doesn't seem so bad. "Sure, it would be nice to walk into a fancy office, but people in fancy offices look stiff and boring, and I never want to be stiff and boring. I want to be fun, so the heck with all this studying."

If the stick isn't bad enough, the hard work may not seem worth it, and you get stuck in the middle again.

I'm explaining this because as a species this is exactly where we are—we're back in school with the need to put in a lot of effort to graduate into something better. If we're not aware of our goals and anti-goals, finding the motivation to do the work is hard, if not impossible. Even if we want to work for a better future for our children, how will we find the energy or the time to do so when we're too busy feeding their hungry bellies?

Right now we might be in the backseat of that self-driving car heading blindly to the outskirts of town with no way back. We don't want to go there, trust me. None of us should because none of us belong there; it's not a world in which we want to live. We belong in a future much more beautiful and abundant than we know today. We *can* manifest this reality if we have courage and willpower to do the work to graduate into a new consciousness and world.

You are enrolled in the class. All you have to do is sign in and commit to the homework. I'm not saying it will be easy. I'm only saying it will be worth it. But my words alone will not create the long-lasting incentive to commit yourself for the long haul. For you to show up, and to keep showing up, you need to *want* to change. That desire can only be birthed from a deeper awareness and by finding a balance between the carrot and the stick. Know the world you don't want, and dare to dream of the world you do want. Then muster the willpower to get yourself out of the snooze car and into the sports car of action.

CHAPTER 19
Two Types of Change

As with most problems we're facing, it's easy to overlook the most obvious. With climate change, we tend to talk a lot about *climate*, but we forget about the second part which is *change*. The most obvious thing we're dealing with in the climate crisis, which can be easily overlooked, is change. Either we don't act, and climate change will continue with unpredictable changes we don't even want to think about, or we act and start *choosing* to change. Either way, there will be change, and the world of tomorrow will look very different.

The issue is that humans aren't great with change. If we were, we would've solved this crisis a long time ago. I'm sure of it. As creatures of habit, we like to keep things just how they are, the way we're used to them being. We feel comfortable; we like the sense of safety that comes from knowing everything is how it's supposed to be. But as we know, that belief is one of the most dangerous things we face, and learning how to adapt to and even *embrace* change is a necessary and heroic action.

How do we get to change?

I've come to realize there are two types of change. One is what I call *fear-based change*. It's the kind of change we accept when it seems like we have no other option, the "fight, flight, or freeze" mode that gets spurred when our safety is at risk. This kind of change is usually quite effective and can trigger instant and massive changes, like evacuating an entire town due to a natural disaster or war. The day before, the people had no intention to move, but new circumstances quickly changed their minds.

Fear-based change is fast, productive, and usually comes with no or very little planning.

The second type of change is what I call *positive incentive change*. This change starts from within and slowly grows into making decisions; it's unforced change that comes from the outside world but is activated from within. You have experienced this kind of change when learning to play an instrument, studying for a test, or getting in shape for the summer. Some sort of incentive on the other end—a goal—kept you motivated to continue acting to move toward the desired change.

Why am I bringing these up? Because, for climate action, it is essential we understand the two to successfully focus on the right kind of change moving forward. As I said earlier, we need to be aware of both the carrot and the stick and find the middle ground where we act early and continue to act.

Let's start with fear-based change.

So far, fear-based change is what we've tried to trigger with our climate warnings. We want people to feel a sense of urgency, like the end is near, and in doing so, upend our lives and take action. The only issue is that for most of us, climate change doesn't evoke those feelings yet. A majority of humans still live with the perception that climate change is distant in either time or space and, therefore, doesn't appear urgent enough to motivate fear-based change.

I'm reiterating here that we don't ignore the necessary changes because we're bad people who don't care. Our brains are simply not wired to respond to this kind of information with action. Unless our house is burning down from a wildfire or is threatened by intense flooding, chances are we will be intimidated by the climate doom, but not quite willing to upend our entire life to "save" ourselves and the planet.

To put it simply, we need two important things for fear-based change to work: 1) an anti-goal (a threat) that feels scary and urgent on a *personal* level, and 2) clear directions or a clear idea of what to do to *avoid* that anti-goal. If our town is threatened by a monsoon and we're told to evacuate, we pack as fast as we can and leave. If someone is pointing a gun at us and our entire body says "run," we run.

With me so far?

Okay, so let's look at climate change. Does climate change feel scary and urgent on a personal level? Again, to most of us, it doesn't, so the sense of urgency needed to act is not there.

Second, do we have a clear idea of what to do to avoid it? Not really. Sure, we should travel less, be more responsible with our shopping, turn off the lights whenever we can, etc. But is there *one* clear action that can launch us into the change needed to get us someplace safe? No. Why? Because we simply don't have one, and because it will take thousands if not millions of small efforts to edge ourselves closer to a better world.

In other words, we lack the two necessary components to make fear-based change effective and sustainable for climate action. Not only that, but for those who are actually personally threated, it is hard if not impossible to affect change. Once you're in survival mode, you're not really equipped to change politics and fuel the transition to a better world.

Not only may threatened people be physically, emotionally, and economically incapable of bringing about change, but something interesting happens in our brains at this stage. Once disaster starts to unravel our lives, and we find ourselves truly afraid, taking bold steps into the unknown is not high on our to-do list.

In *The Influential Mind*, neuroscientist Tali Sharot writes: "Once people feel threatened, they become more focused on the negative and more likely to consider how things can go wrong. As a result, they decide to act in a conservative matter."

Unfortunately, this is already the reality for too many, so it's up to us who still hold the privilege to act courageously to do all in our power to bring about positive change.

Positive Incentive Change

On that note, let's look at the second kind of change. This is completely different in character. Instead of trying to *avoid* an anti-goal with quick action, you slowly work *toward* a goal you're trying to achieve. As mentioned earlier, this kind of change takes a whole different kind of willpower and a clear incentive at the end that inspires you to keep going.

To succeed at this kind of change, you will most likely have to go through a lot of trial and error, adopting a "try, fail, and try again" mindset. Since the nature of this kind of change is slower and requires a lot of work, it can't be forced upon you by outside sources—it is essential to fuel it from within.

Remember what I said about your parents nagging you to study? No matter how much they warn you about having to flip burgers forever, unless you have a vision of a *better* career in mind, it's hard to find the motivation to keep doing all the hard work. This is why you need a desirable goal to work toward.

If we bring it back to climate change, the work we're looking at here is massive. We're not going to solve the climate crisis tomorrow, nor is it going to be an easy fix. It is going to take decades to get this right. Most likely, it will be the work of our lifetime.

How do you remain motivated to do the work? How do you find the motivation to try new things, be ready to fail, and then try again? How do you muster the *willpower* to leave what you know to be safe and put in the necessary work to create the world of tomorrow?

The truth staring us in the face is we need to believe a better world is possible. We need to create a vision so strong it will fuel our work for all the days to come.

Since the fear of a climate disaster is not enough to kick us into action, how can we also begin looking ahead and work for a future *we want*, instead of just trying to avoid what we *don't want*? To hone my point—how do we start choosing change before it's too late and change chooses us?

I also want to add that, of course, it is okay to feel scared and overwhelmed about climate change *and* excited about a new world at the same time. The important thing to remember is not to get stuck in the fear and start looking for a goal to work toward.

In the next section, Awareness and Healing, we will talk about how to turn negative feelings into something empowering, how to funnel them into energy we can use to build something good instead of trying to avoid them.

In the section on Choosing Optimism, you will gain further insight into how important visualization is in taking action and how to continue to fuel optimism in everything you do. Spoiler alert—it's not just about wishful thinking. It involves real science.

In the brilliant book *Climate Courage*, Andreas Karelas (a fellow climate optimist and dear friend) draws from the work of Chip and Dan Heath and their book *Switch* when describing part of climate psychology. One conclusion both the Heath brothers and Karelas come to is that negative emotions tend to have a limiting effect on our brains and fear and anger create a sort of tunnel vision that only allows us to focus on the issue at hand. Since we know working for climate justice is about finding solutions for a new world and *not* fighting the old, it is critical we understand this. And as Karelas concludes: "To solve bigger, more ambiguous problems, we need to encourage open minds, creativity, and hope."

CHAPTER 20
Curtains Can Ruin Christmas

It doesn't take much to ruin Christmas. I learned this when I was eleven and a couple of modern-looking gnomes sought entry into our Swedish home. To this day, I owe one of my biggest life lessons to those gnomes, and they have become an anchor for understanding my own strained relationship with change.

When I was little, I loved Christmas more than anything else, so much in fact that I would start dreaming of Christmas in the middle of summer. I could be on the beach with my family when, suddenly, feelings of Christmas would sneak in under the hot sun and ocean-fresh air. When they did, jolts of excitement spread through my body, and suddenly, I was longing for cold days and dark nights. I wanted to snuggle up in a blanket, light a candle, and wait for magic to come.

Although I've always loved all seasons, I wasn't as obsessed with summer as most Swedes. After a couple of months of beach days and running around barefoot, I was ready for fall. I couldn't wait to curl up with a good book while the rain played music on our windows. I still have a secret love for that weather. Something about rain and cloudy skies makes me feel…excited. Arthur calls it "Tess weather" (Tess is his nickname for me; it's short for Anne Therese), and he's quick to mock me when he gets a chance. "Oh, you'd love it here. The weather is absolutely dreadful."

I think my love for gloomy weather started with my love of Christmas. Because when the weather outside got colder and the days shorter, Christmas was growing closer, and I probably loved the *anticipation* of these magical times more than Christmas itself.

There were rules, though, very important rules. You couldn't listen to Christmas music in the summer—that would be absurd. I also marked my calendar for when it was okay to start lighting candles in my bedroom.

When I got old enough to be allowed real candles, I would have ceremonies with myself where I'd turn all the lights off and only exist in the candlelight. It didn't have to be many candles; a few or a single candle would do. And then I sat there in the dark, thinking, sketching, and writing songs. In those moments, I was connected to what I thought was magic, and I knew it was special. It was the kind of magic that only came out at that time of the year, in times of *anticipation*, and I was careful not to disrespect it.

For each week and each day throughout December, the magic inside me got to grow a little more. The chocolate pieces in my advent calendar disappeared, and the advent candle got shorter and shorter as we lit it every day. It was important not to cheat and get ahead of the dates—rules I made sure my brother and I both followed. I loved every part of counting down the days—magic was coming!

But one year, everything changed. It was a few days before Christmas, and I was waiting for mom to get home from the store. Anticipation had been building up over the past few weeks and we had finally landed on the big finale. It was time to go all in. In our house, that meant gnomes in all shapes and sizes popping up in every corner, and countertops and chandeliers getting dressed in greenery and moss. I couldn't wait.

It was December 21. My magic meter was high, and my spirit-filled body was jumping with excitement. I couldn't wait to transform our home into the Christmas winter wonderland it became each year.

My favorite part was waking up early Christmas morning before everyone else. First, I would go into the kitchen and light the stars in the windows. Because of the specialty red and gold Christmas curtains Mom would hang every year, the stars would shine a warm light across the room. That's where my Christmas morning magic began. After that, it was time to wake up my little brother. We would unhook our stockings, which had been filled with packages in the night, from the doorknobs and carry them into the living room where we would sit under the tree and open them quietly while Mom and Dad were still asleep.

Those mornings with my brother were so special I could long for them in the middle of the summer. Those were the moments I dreamt of all year. I wouldn't let anything ruin those moments, and I honestly thought nothing could. But I was wrong.

Mom walked through the door with a handful of shopping bags she dumped on the hallway floor. She was in good spirits, and like me, she couldn't wait to get started.

"I'm so excited," she said, "and I have something special this year!" She unwrapped her scarf and hung her coat on the rack, all while I was standing by, anxiously waiting. I wondered what it could be as my anticipation grew—I loved surprises.

"Really, what is it?"

"Just hang on. I'll show you."

I waited patiently as she took off her boots and put her things in order. The shopping bags on the floor were screaming for my attention. When she'd finally undressed, she turned to me with a smile.

"Look what I found!" She grabbed one of the bags and walked it over to our dining room table. "I thought it was time for new curtains in the kitchen; don't you agree? The red ones we have are so old, and they make the place look so dark. I found these in the store and absolutely fell in love. What do you think; aren't they adorable?"

Mom was so caught up in her excitement that she didn't notice what happened to me. Because right there at her side, my world froze. In absolute shock, I watched as she pulled the curtains out of their plastic cover and laid them out on the table. They were…beige. Thin. Straight. And even worse—they had some modern-looking gnomes all over them. They were nothing like the velvety red, fluffy curtains that made the kitchen look so magical on Christmas morning.

Mom turned around to ask for my approval. "What do you think?"

What did I *think*? *They're terrible*, that's what I thought. I couldn't believe she would do this. How could she even think of ruining Christmas like this? Didn't she know this holiday was all about traditions? Didn't she know how much I had waited for Christmas Eve to come? For months I had counted the days and dreamt about it. I'd dreamt about how I would feel when I tiptoed out of my room in my pajamas and into the kitchen to light those stars. And now she was willing to ruin all of that because she wanted new curtains?

I remember my response was unreasonable. I cried, screamed even, and ran to my room where I slammed my door. I fell to the floor and wallowed in despair as I felt the magic rapidly rush away.

Everything. Was. Ruined.

You probably think I was absolutely ridiculous. A spoiled brat who should learn some manners. It's okay to think that way. I would too, but I look back at that little girl today with so much love and compassion. The eleven-year-old me had yet to understand the importance of embracing change, and it would be many years before I came to that realization.

CHAPTER 21
Fear of Change

Luckily, Mom came to her senses and decided not to replace the kitchen magic that year. After a couple of hours of fussing and crying, I crawled out of my den (my bedroom), and we continued with the decorations. However, a year later, *I* came to my senses after having contemplated that we could probably make some magic with new input, and Mom finally got to hang her new curtains. The modern-looking gnomes have made it up every year since then, and I love them to this day.

But why on Earth did I make such a big deal about curtains? I know this may seem silly, and I get it if you think I was the biggest brat in history, but what it all really came down to was fear of change.

When I was younger, I didn't do well with change. In fact, I wanted things to be exactly how I expected them to be all the time. I found tremendous joy in envisioning my near future, and I almost lived more in the next moment than in the now. Reminiscing about past happy moments, I wanted to recreate them the best I could and thought that was the key to happiness—to recognize what had made me happy in the past and use that as my formula for creating my life moving forward. If I could assemble the right elements to recreate the moment I remembered so dearly, shouldn't I be able to feel the same as I did then?

I loved holidays because, to a certain extent, they tend to be the same. They're infused with traditions and rituals, and the whole point is to recreate them again and again, to make them appear the same, year after year, always with the same happy outcome.

Over the years, I've come to understand so much in our world and society is built on this idea of recognition and trust. Food, culture, customs, and fashion—we want to know what to expect, and we go to great lengths to ensure those systems are kept in place. Therefore, we side-eye those who do their own thing and feel hurt or insecure when someone close to us decides to change. If someone else changes, it means we could change too, which invites questions we may not be ready to grapple with. Our fear of change, I've learned, is one of the biggest things holding us back.

When I first decided to go vegan in my early twenties, my grandmother, the woman I had been the closest to (except for my mother) since birth, didn't speak to me for a while. She was hurt that I would no longer eat her food—the one way she had always showed her love for me—and was scared of these new choices I'd made and the effects they would have on my health.

At first, I was hurt too. I couldn't understand why she acted this way. Didn't she see I was doing it for the animals and to help save the planet? Didn't she see I had made this decision out of love? But then I understood that, of course, she didn't see it that way because a vegan or vegetarian diet was not something she was familiar with. To her, having grown up in a dairy and meat-heavy country like Sweden, it was a foreign concept that held a lot of hidden threats to the grandchild she loved so much.

When I recognized this, I knew the only way to mend our now somewhat broken relationship was to invite her into this new life of mine. I started making trips to my grandma and grandpa's house with ingredients I had bought, and we would cook a delicious, plant-based meal together. I wanted to show her that I was still eating, and I wasn't missing out at all—either on nutrition or the pleasures of food. We had some great times cooking together in that kitchen, and our relationship was soon mended. On top of that, I got to introduce my grandparents to new recipes and new ways of cooking that are better for the body and the earth.

Today, my grandmother understands and respects my choices, but I also say yes to a cookie or two when she offers her love because I've learned that the shift can only take place if we meet each other halfway. We must recognize we come from different worlds and that the world in the middle is still messy, filled with unanswered questions and steps yet to be taken.

I'm sharing these stories, and especially the Christmas curtain one, to show you how resistant I used to be to change and how much I understand anyone who's struggling with embracing any sort of new. In the past, I was not a master of change (quite the opposite), but I have been through a lot of changes, both deliberate and unchosen, that have truly shown me the beauty of living life always curious about the next answer.

I've been wrong more times than I've been right, and these days, I find being wrong fascinating—it means I get to change my worldview once again and learn something new. It means I get to continue to *grow*.

> "The measure of intelligence is the ability to change."
> — Albert Einstein

We will talk more about the beauty of being wrong later, but for now, check in with yourself to see how open you are to change. Do you have a story similar to my Christmas curtain one? Do you feel hurt by someone in your inner circle who decided to explore new ground? Are there changes you wish you'd made a long time ago, but have yet to find the courage to act on? Below are a few journaling prompts I encourage you to use to gain a better understanding of your own relationship with change. Next up, we'll dive into how to improve that relationship.

Journaling Prompts on Relationship to Change

1. On a scale from one to ten, how open are you to change in general? (How often do you try new foods, visit new cities, or talk to people you don't know?)
2. Is there anything you always wished to try/do that you have yet to find the courage for?
3. How set are you on your current identity, and how much pride do you take in the person you see yourself as today? How offended do you feel when people "attack" this identity of yours?
4. How open are you to listening to and accepting other people's viewpoints and opinions?
5. How different would your life look if you were more open to change?

CHAPTER 22
Complain a Little, Then Get Used to It

It is amazing how good we are at fussing over things. Really. One of my favorites is when we complain about the Wi-Fi being bad on airplanes. Like, I get it; you expect it to work. But I mean *come on*—we're traveling through the *air* and we're fussing over the *internet* being slow?

In moments like these, I like to catch myself and reflect for a second, asking myself how our expectations get in the way of the bigger picture. How much fuss is there because we've gotten used to expecting something, and how easily could we get over that fussing and get used to something new?

If you're an airplane Wi-Fi complainer, I see you. But hey, I feel you too. It's annoying, isn't it? If you pay for something, you want it to work. Even if you *don't* pay for it, you want it to work because you were told it was going to work and now it does not. That leads to mega-frustration.

But isn't it fascinating how fast we get used to new expectations (I should be able to send this email 30,000 feet up in the air) and how easily we get set off course when those expectations aren't met? Even more fascinating is recognizing how quickly we move on once we stop fussing.

Humans are creatures of habit. Not only that, but we love to *defend* our habits. It's because we tend to give our habits a sense of identity and feel like they're not just what we *do* but who we *are*. When we deliberately act in a certain way, we're *doing*. When we go about our daily actions without paying much attention to what we do or why, we're *being*.

Of course, this is not true. Habits are also an act of doing, but since we rarely pay attention to them, they start to become integrated into the core of who we are.

I can't stop eating meat—I'm a meat-eater; it's who I am.

On our second date, my now-husband Arthur told me he couldn't stop eating meat because he was a man. Somehow, I saw through his bull and stuck around. (Thank God I did. He's a true gem.) Not that I can't date someone who refuses to stop eating meat (although seriously *cutting down* is what we should all be doing for the sake of our planet and the welfare of animals across the world), but because I don't think I could've married someone who put such limitations on himself. Would he not be a man anymore if he quit eating meat? What if the world suddenly were to run out of meat? Would our species simply die out?

Of course, he knew deep down that he was not a man because he eats meat, and he continued to earn my love by showcasing, over and over again, that rethinking manhood is the manliest (and in my opinion, one of the sexiest) things you can do. I love to cook, and he always eats what I cook (which I would say is 97 percent plant-based). Besides, we've had some of our most romantic dates at vegan restaurants, and he's always down for trying something new.

My point isn't that cutting meat out of your diet is the answer to everything. My point is that we tend to forget the habits we cling to so dearly are nothing but a repetition of doing, and with any doing, there's an active choice available to continue doing what we've always done or to try something new. Our egos can survive some change. I might even go so far as to say it will strongly benefit from constantly questioning what we do and why we do it.

Sure, there will be some fussing, even when no one is around to complain to. You can fuss to yourself too, and you most likely *will* fuss, so be prepared for it. Doing something new is always hard because you're not *used* to it. But you can *get* used to it, and once you are, that will be the new easy. You will have created a new habit, which will stick with you until you again choose to question and try something new.

I love it when I hear people say, "I could never live without [blank]." It could be pizza, cheese, wine, chocolate, coffee, etc. (Okay, I actually thought coffee was impossible to live without until I got pregnant, and suddenly, I couldn't stand the smell of it anymore.) People think they can't live without their favorite foods because they love them too much, *crave them* too much.

But do you know *why* you crave a given food? Because your body is used to it, that's why. And because you have a bunch of gut bacteria in your belly that are used to eating it and crave it too.

You may not know this, but about half of your body is made up of foreign bacteria, and they run a really funny show in there. Not only do they help determine what your tastebuds prefer, but they also influence how you sleep, feel, whom you attract, your mood, and all sorts of shit. If you find this fascinating, wait until you learn more.[7]

It is important to recognize that almost nothing about you is set in stone. It's not because of DNA you were born with and can't shake, nor is it about some sort of unfortunate fate you got stuck with. Who you are is simply a combination of the things you do and the daily habits you've gotten used to doing.

What's truly fascinating with us humans is we can get unused to things and used to *new* things pretty fast. I know this personally because I don't eat so many foods anymore, and I don't feel like my life has ended, nor do I feel like I'm missing out.

I can't eat pizza or sandwiches because I'm highly sensitive to gluten. I can't eat dairy because I'm allergic to milk protein and lactose (unless it's from sheep or goats). I can't eat a lot of sugar because I'm pre-diabetic. I also don't eat a lot of these foods because of my ethical beliefs and concerns about the environment, so I *choose* not to eat much chicken, fish, dairy, or meat.

But don't worry. I *can* eat so many things, and I choose to focus on those delicious foods instead.

Sure, it took some serious determination to change my food habits. Before I began transitioning to a plant-based diet, I avoided carbs like my life depended on it, and I filled my plates to the brim with meat, chicken, poultry, and fish. An avid gym junkie, I believed protein was gold in my veins and the safest source of protein was animals. However, after a summer of reflection and a couple of documentaries, my worldview changed forever. Having seen *Cowspiracy* and *Forks over Knives*, and having read *Eating Animals*, and spending hours working with a guy who kept asking me why I was so set on consuming other life, my thoughts about food

---

[7] If you're interested in diving into gut health and microbiome, a truly fascinating topic, I highly recommend checking out Dr. Gundry's books, *The Plant Paradox* and *The Longevity Paradox*.

were seriously changed. I didn't want to eat animals anymore and knew my diet had to undergo a transformation.

To be honest, I was terrified. I worried I wouldn't get the protein I needed, and that my health would quickly deteriorate. I had to spend a lot of my hard-earned money hiring a dietician to help me. I needed someone who could assure me if I replaced the animal-based protein with plant-based sources, I would be fine—better than fine. But I didn't believe him at first. I was *fussing*—big time. The only way to stop that fussing was slowly but steadily to incorporate new habits. Before I knew it, I got used to something new.

What struck me the most was how *fast* I adapted and learned to love new things. I began with creating a few new meals I added to weekly meal plans, and soon enough, I began craving those meals instead. When I walked home from the gym in the morning, I no longer craved the yogurt and granola I used to eat—I craved my overnight chia pudding with crushed nuts and blueberries instead.

The magic I was introduced to through this whole transition was that our bodies are actually much more capable of change than we give them credit for. Moreover, they will *reward* us for making choices that are good for us.

If you start eating healthier, you will definitely feel like you're punishing yourself at first, but very soon, your gut bacteria will be replaced with the kind of bacteria that *want* you to eat healthily and thrive on foods that are good for you.

With the right bacteria in your belly, these new gut buddies will start operation "clean house" and begin to introduce an environment in which *good* bacteria can thrive. With those good bacteria in your system, you will start craving healthy foods instead, foods both you and the bacteria thrive on. Quite fascinating, isn't it? Your tastebuds will change, too, and you'll much prefer a crispy, fresh salad over greasy, salty fries. I know it's hard to believe, but it's true.

If you've never tried changing your diet, I challenge you right now to give it a try. It's quite fascinating, and you'll gain a serious understanding of the power you hold to change your life at any given time. Just don't give up too early. Make sure you stick to your commitment long enough for your gut bacteria to change and for those new cravings to kick in. From there, moving forward will be a dance of better health, restored energy, and even more so, some serious self-discovery.

Sure, there are those few occasions when I really wish I could have a slice of pizza. I think it's unfair that Arthur can eat so much "crap" and be absolutely fine. I wish from time to time I could eat crap too and not have to worry about it. But the truth is 99 percent of the time, that crap will make me feel like shit, and I just don't want to feel like shit, period. *Treating myself* is not a treat when I have to pay for it later, wouldn't you agree?

Besides, I do treat myself. There are many things I indulge in like over-priced oat-milk Americanos at a cute coffee shop, dark chocolate with sea salt and almonds, hearty salads, and luxury oats (what we call weekend oatmeal with all the right toppings). None of these make me feel like crap later, so they feel much more like a treat than any gluten in the world would for me.

But I must confess right now that I, too, fuss sometimes. I like to make it clear to the world (read my husband) that I'm "missing out" not being able to eat pizza or the delicious-looking pasta dish, just so he can recognize my sacrifice. However, once I stop complaining, I remember I'm actually better off with the choices I've consciously made for myself, and I stop fussing.

And life gets so much better once you stop fussing.

I want to help you get better at quitting the fussing so you can begin to question the code of your being and continuously grow into someone new. Don't underestimate the simplicity of changing your diet. It might seem like it's a choice that will only benefit you and perhaps some animals down the food chain, but don't be mistaken—a transformation is taking form that you may not be able to see, but when you make small changes to your everyday life, your identity will start to change, and the world around you will to.

We ask for systemic change, not climate change. We demand shifts that can transport us into different—better, fairer, and more sustainable—worlds. Now, we must begin to recognize that we hold the power to make those shifts. We *are* the bodies that can activate those changes from within. We need political and economic changes, yes. We also need radical changes to our justice system and how we evaluate every single life on this beautiful Earth. But we also need big changes in our consciousness, a rebirth almost, where we get to question why we're here and find our way back home.

That change can only start with you. It's an ongoing journey that only you can accept and fall in love with, and it can start right now.

You, my friend, are a force to be reckoned with. Every morning, you are given the choice to actively keep doing what you've always done and continue to feed into the world you've always known, or to choose again. The second choice is a scary one, but it's also empowering beyond belief, and you will soon recognize that.

If you decided to become and embody a vessel for positive change, there really is no stopping you. If you want to see change in the world and discover what a life of happiness, health, and purpose looks like, this is the power I'm talking about right here—the power to find the courage to choose change.

It doesn't have to be a mountain a day. Start small, set yourself in motion, and you will soon see those mountains growing closer and more surmountable by the day. Many quotes confirm just this. Many songs, stories, and movies tell us tales of how small, individual forces, when brought together, become a power so big new worlds take form.

One of the world's most famous quotes is by anthropologist, author, and speaker Margaret Mead, a quote that's been edged in my memories since I first read it:

> *"Never doubt that a small group of thoughtful, committed citizens can change the world; indeed, it's the only thing that ever has."*

You will be amazed by the power you have to change not just your life, but the world once you learn to master the skills I'm about to teach you.

We are not meant to stay the same. That is the truth I can swear by, so buckle up and get ready for some serious movement. Things are about to get exciting.

CHAPTER 23
Being Wrong

Sometimes I wonder how much pain and suffering exists in the world simply because we don't know how to be wrong. How much harm could have been avoided had we just been better at letting go and thinking again?

We don't like to be wrong because society has taught us to feel like fools if we are. We are supposed to be smart and educated, to know where we stand on certain topics, and to be willing to fight for our opinions when they're questioned. And although there are definitely times when we must fight for what we believe in, the *status quo* of wrong equals weak doesn't help much when it comes to creating change. If anything, it holds us back, making us miss opportunities to move forward. To embrace change, we must start by redefining our relationship with being wrong.

Not only does it take courage to accept you're wrong, but adapting this open mindset will invite you into a whole new world of possibilities. You learn it's okay to be wrong because only when you're wrong can you learn something new and do better. And as you leave room for failure, you also leave room for improvement, and that's where the innovation is sparked.

As I write this, I'm putting myself center stage. I'm seeing myself in that spotlight, and I'm saying "Yeah, lady, that's you."

I have been wrong an embarrassing number of times, and I've preached my wrongness with passion, the kind of passion where one goes, "Listen, I've found *the* way (to live life, lose weight, save the world, etc.). Now you better pay close attention so you can become enlightened too."

Why does someone preach a message with passion? Why do they hope others will get onboard as well? Why as a teenager did I try to convince all my friends the diet I was currently on was the answer to being happy and that they better trust what I said? Because I believed in it. And because I cared deeply and wanted to share what I believed.

Turns out I'm not alone in this. As a species, we're somewhat addicted to sharing information. Tali Sharot, a professor of cognitive neuroscience, writes in *The Influential Mind* about a research study conducted at Harvard University. The study showed that when people had the opportunity to communicate pearls of wisdom to others, their brain's reward center was activated. We experience a burst of pleasure when we share our thoughts. Maybe that explains why, as Sharot shares in the book, "Every single day, four million new blogs are written, eighty million new Instagram photos are uploaded, and 616 million new tweets are released into cyberspace. That's 7,130 tweets per second!

Rest assured, there's no lack of information in our world. Or opinions for that matter. I read somewhere many years ago that we have access to more information in one issue of the *New York Times* today than most people in Shakespeare's time were exposed to in their lifetime. But what this information is, more than anything, is a mirror of our current reality. Millions, if not billions, of mirrors, reflecting back to us what is true and what we ought to think is true right now. But if we're so busy talking about what the world looks like right now, how will we find time to envision a new one? And if we can't envision a new one—if we can't find it in us to believe that things could get better—how will we ever find the courage to choose change?

People who are wrong are not bad people. It's not like they stay up late at night and scheme away, making up all the lies they can introduce into the world. People who are wrong are not liars.

People who are wrong are also not ignorant or dumb. It's not like they pick up some values off the streets and then hold on to them forever without ever questioning them. People who are wrong are not lazy.

In fact, the most intelligent and well-educated people tend to be the ones who are "wrong" the most. This has nothing to do with intellect and everything to do with how much we've invested into believing something.

If you spend years of your time and a lot of your valued dollars on a degree of any sort, you will hold onto what you've learned during those years with tight fists and pride. The greater the investment in information, the harder it will be—psychologically—to let go of what you've learned.

Adam Grant, popular science author, and professor at the Wharton School of the University of Pennsylvania specializing in organizational psychology, wrote a whole book on this. *Think Again: The Power of Knowing What You Don't Know* is a true masterpiece filled with insights and knowledge from big thinkers of our time. It truly shows that we win by staying curious. Grant even goes as far as suggesting we ought to fall in love with being wrong; we should be excited when our ideas are incorrect because it means we get to learn something new yet again.

Scientists are well aware of this feeling. Their entire job is built on questioning the *status quo* and learning something new. To find something that contradicts their original belief is exhilarating because it means a breakthrough of some sort, a gateway into a new world of knowledge. We ought to treat information and perceptions the same way. How can we fall in love with the idea that everything we think and know today might not be the best way to move forward? How can we stay curious in exploring new and better ideas? What if the world of tomorrow isn't just different but *better?*

Here's a thought to chew on—you are allowed to change. Just because you grew up believing one thing or learned something in school does not mean you have to believe in these things forever, nor do you have to agree with what the person you were back then thought or felt. You are allowed to *change,* which is also why it is absolutely okay to be wrong.

In fact, we should strive to get really good at being wrong as often as we can.

"It's a rare person who wants to hear what he doesn't want to hear."
— *Attributed to Dick Cavett (quoted in* Think Again *by Adam Grant)*

We have to realize the Earth needs us to be wrong. Mother Gaia is holding out her forgiving hand and saying, "Dear children, I see your efforts. I know you want to do great things. But sometimes life's gotta teach you the right way, and it can only do so if you let it. I forgive you for your mistakes, and I ask you to please think again."

We are not dumb, ignorant, or evil for being wrong; we're simply people. People make mistakes. People learn new things that challenge their current beliefs and worldview. People have experiences that help them *grow*.

Why don't we let ourselves be fully human? Why don't we give ourselves a break and remind ourselves it's okay to think again? If we are to find our way out of this climate chaos we've created for ourselves, we need a lot of bold rethinking. We need to accept that how we built our society up to this point isn't quite working and we need to go back to the drawing board. We need to learn from our mistakes and find healthier, better, and more sustainable ways moving forward. This doesn't mean we have to sacrifice all the things we love and reverse our evolution. On the contrary, we must *redefine* what we value and continue to evolve into even more sentient beings. We have to come up with even cooler and smarter things than we have today. Change does not mean giving up. Change means looking at the current situation with critical eyes and giving ourselves permission to think again.

I believe we can rise above our stubborn nature and show we are, in fact, beings of greatness—a species capable of change and evolution unlike anything we've ever seen. And we can do this, not just because the environment forces us to, but because we decided to *take control* of our evolution and actively choose change—over and over and over again.

You have the key to a revolution right now. The question is, are you brave enough to use it?

Keep in mind that trying to force change on someone else never works. People live and abide by their own beliefs, and the more you tell them they're wrong, the tighter they hold on. An attack on someone's truth is ultimately a threat to their perceived safety and their internal alarm system will start going off. (More on this in the next chapter.) Yours will too if anyone attacks your beliefs.

That is why we can't change other people; they can only change themselves. But we can nudge and inspire others to get to the place of *wanting to change,* and that is ultimately what being a *leader* is all about. Leaders don't tell others how to think and feel; they inspire others to become leaders in their own lives and activate their own power from within. Only then can others continue to grow and add more strength, courage, and brilliance to the world. Only then can others learn new things, become new leaders, and pave the paths for better tomorrows.

And not only will those who follow you learn to accompany you, but they will also challenge you, enabling your further growth as well.

Learning how to be wrong is brave, the trait of a true hero. I know that acknowledging you're wrong, if even to yourself, can be hard. It will often feel uncomfortable, especially until you get used to this new way of thinking. You will have to practice, but you will get better over time. As you do, you will not only begin to feel excited about the possibilities of new realities and worlds, but you will also feel free—free of old limiting beliefs, repeated thought patterns, of "this is how things should be" thinking that is keeping us stuck in the current loop.

We *can* create a different world. We have the science, the technology, and the knowledge to reverse global warming and build an even more beautiful existence on Earth. The only thing we're missing now is the willingness to think again, to let go of what we currently believe, and aim for an even better world. That will take courage and many humbling experiences, but as we get better and better at being wrong, a new reality will slowly unfold all around us.

We have the key to unlocking so much positive change. We don't need to wait around for the answers. Many of those answers are already here. We just need the courage to see them. We need to take a good look at our relationship with change, and that starts with building our ability to be wrong, no matter how small and insignificant it may seem.

You don't have to be wrong all the time, of course. Sometimes new information will just reinforce your current belief. Perhaps sticking to your ideas in those situations is important because you are the one who actually has the new, valuable information that can lead others into the new world.

Navigating this landscape can be tricky, and the only way to get better at knowing when to change your ways is to be tuned into yourself. What feels right to *you*? What is *your* intuition telling you is correct at this very moment? Learning how to trust yourself is critical on this journey. (We will talk a lot more about this in the next section.) In our next chapter, I will share an exercise I call "retruthing" that will help you build your being wrong muscle and get better and better at embracing change.

CHAPTER 24
Retruthing

Once we get over the fussing and stop complaining, we can get into retruthing. If you were about to reach for your phone to google the meaning of retruthing, save yourself the time—I made it up. As a Swede, I see it as one of my superpowers to every now and then fill my sentences with words I think make sense. Usually, a friendly soul will correct me and tell me the proper word, but this one I'm keeping because I just don't know a better way to explain this idea.

Retruthing
(Retruth-ing, i.e. redefining truth)
Verb

The willingness to question what is and to let go of ideas, thoughts, and perceptions as they have lived in our heads up until now to create a sustainable and more compassionate world.

Retruthing refers to one's ability to understand things and circumstances always change, and it is our duty as humans to adapt to these changes, to constantly find ourselves in new worlds, communities, and realities—new truths.

Put simply, retruthing is the act of questioning our "truths." Not scientific truths like the ones Einstein and Newton figured out. (Unless you feel tempted to challenge the physics of everything, because if so, go for it.) The truths I'm talking about are the ones we live by day in and day out without really paying attention. I'm talking about the beliefs that shape our thinking and habits, the coded network of thoughts and ideas that make up the script of who we are.

The reason we fuss about change is we have, in one way or another, been brought outside the script of our comfort zone. We're dipping our toes in waters beyond our truths. This will always be uncomfortable because when we act outside of our script, or when the world doesn't seem to fit what our script tells us, it feels a little icky. More often than not, this discomfort shows up as fear that holds us back.

To overcome fear, all you have to do is the very thing that scares you, and in doing so, change a part of your code, which is, of course, easier said than done.

I started a business in my late twenties under the name Role Models Management. It's an ethical modeling agency representing models, actors, athletes, and other talent who care about people, animals, and the planet. Our models are not just models; they are Role Models.

I got the idea of the agency when I was modeling in New York, but I lacked the platform I wanted so I could speak up about things I believed in. I wanted to showcase ethical and sustainable brands, not pretend to be a mannequin at Bloomingdale's or spend hours on photo shoots for clothes I knew were polluting the Earth. I figured other models out there must feel the same way and long to work with brands with similar missions and values. I shared my idea with a few people, and later that year, I launched the agency with a woman I had connected with who had similar thoughts.

Val Emanuel and I had never met in real life. She was in Los Angeles, and I split my time between Sweden and New York. We were both struggling artists (read broke) and didn't have much money, but what little we had we put into starting the business. This meant we acted on trust and launched a business with joint ownership, having never met in real life. I often marvel at this story and the guts it took to make it happen, but a deep sense of trust was present from the very start. It turned out to be a brilliant idea that gave birth not just to a business, but a network of Role Models coming together to make the world a better place. I also found an incredible friend and sister in Val, and when we finally met, it was soul-sister-love at first sight.

However, our cute origin story is not the point I want to make here. What I want to share is how undesirably uncomfortable those first months (if not years) of running the business were. I had no experience running an agency or managing talent. I had only been modeling for a few months myself and had very little experience in the industry. I was fairly new to New York, *very* new to the modeling industry, and had, in all honesty, little idea what the heck I was doing most of the time.

I'm good at thinking up new ideas. I'm excellent at shaping a brand around that idea and launching something that looks cool and visually appealing. I come from marketing and have a reflective mind tuned to think up groundbreaking ideas. I am what some would call a visionary.

Luckily, my partner Val had a bit more experience in the game, having modeled and acted since she was two and even managed a few model friends from time to time. She understood casting directors, producers, callbacks, usage rights, agency fees, etc. She was the business in our business. However, she wasn't okay with me sitting in my boyfriend's Brooklyn apartment and building the website all day. She needed me to dip my toes in the deep water and start making some business appear.

That money ain't gonna come by itself, mama.

She was right, but the very idea of having to interact with *humans* terrified me. So much, in fact, that when I got a voicemail from a guy who introduced himself as a casting director calling with some questions he had regarding someone we had submitted, I freaked out. "Call me when you receive this," his message ended.

I didn't. Instead, I called Val.

"Val, I have a voicemail from a casting director asking me to call him back. What do I do?"

There was a short dead silence on the other end of the line. Then she simply replied, "Call him back."

Things like this happened a lot in the early days of our business, and I hated myself for even having started it in the first place. I couldn't understand why I would put myself in such an uncomfortable position. I owe it to Val, an Afro-Latinx woman who simply takes no bull, for forcing me outside my comfort zone. She didn't ask me to please go feel the water; she pushed me right over the edge and watched me land with a big (and not so graceful) splash in the pool of spiritual growth.

I have left a lot of comfort zones in my life, and in doing so, done a lot of recoding my script. How I see the world today is completely different from how I saw it ten years ago, and how I find myself and my place in it has shifted dramatically as well.

Having understood and fallen in love with the act of questioning, retruthing, and *growing*, I have entered a state of always becoming someone new. It is clear to me we are not meant to stay the same. We're meant to *grow*. How do I know that? Because it feels *amazing*, and not once have I looked back and thought, *Dang, I wish I hadn't learned this lesson.*

Sure, life was simpler as a kid, but I would not for a second trade the courage, strength, humility, and wisdom I've gained over the years to go back to simpler times. And I'm sure you feel the same. Reflect on who you are today and how much you differ from the person you were only five or ten years ago. Quite a difference, isn't it?

Reflect for a second. What life lessons and challenges molded you into the person you are today? Are you proud of this version? I bet you are. Because regardless of mistakes you might regret or difficult times you'd rather forget, you are who you are because of the journey life brought you on up until now. This is truly spectacular. This means that not only are you a conglomeration of molecules and DNA unlike any other body on Earth, but your personal journey is unlike anyone else's too. You are unique, so don't forget to celebrate that more often.

Recognizing you are the result of your past experiences, aren't you curious to find out who you can *become* in the years to come? Isn't it exciting to think about all the versions of yourself you haven't met yet, and how stunned and proud you might be of these new versions? And if you know experiences that challenge your beliefs and thinking make you grow, what happens if you actively begin questioning and choosing growth? Can you speed up the process of meeting these other versions of yourself?

Yes, you can, and that, my friend, is exactly what *retruthing* is all about. And don't worry; it's not like you have a limited set of future yous, and if you speed up the process, you will reach your maximum growth faster. Who you can become has no limits, and the sooner you get going, the better your chance to reach versions of yourself you never thought possible.

This world needs a lot of things, but more than anything, it needs curious and courageous souls who are not afraid to question themselves and grow. With growth comes new mindsets, new curiosity, new visions, and new ideas. When we grow, we make it possible for things to get even *better*—technology, housing, work hours, food systems, community structures, etc. The world we know today is nothing but a bunch of realities we have conformed to that can always be questioned and changed.

You don't have to start by questioning the system; you can start by questioning *yourself*. When you grow, your vision for the world grows with you, and the system will soon follow. This has happened throughout time. The only difference is that now we need this transition to happen faster. We need big, bold changes, and we need them now.

We need heroes to help us make the change. You can become a hero by working toward empowerment and rapid growth—which come with *retruthing*.

CHAPTER 25
A Coded Computer

To understand the concept of retruthing, you must begin by understanding you're much more like the minicomputer in your hand than you might think. Just like your phone can open apps, play music, and send text messages because of a bunch of code, you run on code too. The only difference is you don't get a notification when it's time for an upgrade. You don't get an alert saying, "A new version is available. Are you ready to update?"

So how do you know you're operating on the latest, most optimum version?

When you get a new phone, it works seamlessly. It's fast, clean, and has all the latest updates. Like a shiny new toy, the new phone works like magic. But after a few months, your phone begins to fill up with photos, apps, texts, and failed TikTok videos. It starts to feel slower and less magical and you might be getting notifications that you need to download the latest update. Your system is too old, and a new and better version is available.

What do you do? You'll probably delete some bad photos or duplicates, old messages, and apps you're not using. And per request from your minicomputer super-friend, you kindly plug it into the wall and install the latest version of software. Like a good custodian of your metallic friend, you do what's needed to keep it fresh and up to date.

But how about you? How often do you go into your internal storage to delete, clean up, and update *yourself*? Believe it or not, you are just like a smartphone. You start with a clean slate, and as time goes by, experiences build up and code your mind and personality with all sorts of information. Whether you're aware of it or not, your subconscious code and operating system is constantly being altered and rewritten, and unless you're aware of this, your software can start to take on a life of its own.

How do you know you're not operating on old data? How do you know if who you are is the person you'd like to be now or the person who got coded years ago? And if you don't know the answer to that, how can you be sure who you are is someone you're proud of, or if you're even living the life you want to live?

These questions hit me like acorns from an oak on a fall day when it became clear most of what I do has very little to do with my conscious choices, and a lot to do with the coded script I am barely aware of—a script that, unless I pay attention to it, will keep sending me in an unknown, unplanned direction I may not even want to go in. How do I know if the choices I'm making every day are actually edging me closer to my goals if I barely know I'm making them? What if—and this is scary—I'm the one *stopping* myself from growing?

Coming to this chilling realization, I came up with the practice of retruthing. I wanted to find a way to empower myself and ensure I keep questioning what I believe and what I *want* to believe. If I don't, I'll never change, and the world around me will either run away from me or stay stubbornly stuck in old ways when it could be moving forward.

I don't want that. I don't want to be left behind, and I don't want to be the reason we're stuck. I want *change*. I want it so badly I get antsy and quite hard to be around if things move too slowly. If my hubby spends too long in the cereal section at the grocery store trying to pick the right kind, I get all complain-y and make a fuss about how the day is slipping through our fingers. *Just pick one already.*

But it happens with bigger things too. I want the world to change. I can't wait to see the future we will find ourselves in once we've learned to let go of the world we know and build something new, something even *better*. What will it be like to walk down the street without the noise and smell of cars? That's just one of the many visions I like to ponder.

To enact this change, I need to be the change. And to be the change, I need to understand the code in place that wants me to stay the same. That's where retruthing comes in. It's one of the most powerful exercises you'll ever do—one that unlocks doors to your subconscious, rewrites the script of your life, and paves new paths to whole new worlds.

Let me explain how it works.

A Coded Script

Ever since you were born, your code has been written using a ton of data that makes up the living, thinking, and feeling person you are today. Past experiences, culture, family, friends, and anything that happened to you before today helped program the code that is now your "truth."

It's easy to think it's everyone's truth, but it's actually only yours. Your truth will inevitably differ from other people's. Some will be closer to yours than others. Inside your family, for example, many of your truths will be shared. But since you have your own friends and experiences, life events no one else has gone through but you, your truths are uniquely yours.

The truths I'm talking about are coded into your subconscious. The subconscious is like your own personal lifeguard who's there to pay attention to your surroundings and warn you when something seems off. It exists so you can go on living without having to spend energy rethinking everything you've already learned—how to breathe, walk, eat a sandwich, say "I'm sorry" when you bump into someone on the street, those sorts of things. The majority of what you do, from how you put on your shoes to which salad you order for lunch, is determined by the coded script in your subconscious.

We need our subconscious or we'd be spending our days figuring out how to open doors and operate a microwave, which would be extremely time-consuming and annoying. Our subconscious also helps keep us safe by signaling a red light means stop and wait, or a group of shadows in a dark alley means we should turn around and walk the other way. All those little things—gut feelings, habits, triggered responses—come from your code. You act a certain way because at some point that behavior was added to your code.

But it gets trickier. If you're told as a child that people of color are dangerous, your subconscious mind will automatically warn you when a person of color enters the room. No matter how anti-racist you want to be, if that code is strong enough and hasn't yet been questioned, that's how you will respond. If you've grown up believing money is hard to come by, you might find yourself struggling with money. No affirmations or mantras in the world will help you overcome these behaviors unless you tap into your code and change it first.

What if you're still doing things out of habit that you were taught as a kid? If you're about to turn thirty, that means you're operating on data that is twenty-plus years old. *Talk about outdated software.*

Bruce Lipton is an extremely funny guy. He's also the father of epigenetics, a new field of study that has gained attention in recent years. Epigenetics is the study of how your behavior and environment can cause changes to your genes, claiming you're not a "victim" of your DNA, but in fact, you are very much in control of changing it. You can change your genes—and therefore also your faith—at any given time. Bruce Lipton speaks about this with passion, and I highly recommend tuning into the interview my cohost Robin and I had with him on *Hey Change Podcast* (Episode 69 and 70) to learn more. He is, if you ask me, a champion at retruthing.

What's mind-blowing about the studies of epigenetics (and if you've started to drift, come back to focus because this is important) is that you spend 95 percent of your days in your subconscious! That means for the majority of your waking hours, you're operating from a script you barely recognize is there. The million-dollar question remains—are you living your life or someone else's?

Retruthing is the act of questioning your truths. If your truth tells you since your mother, grandmother, and great-grandmother all had breast cancer, it's in your DNA and you may get it as well, it is simply a belief you live by. A belief, however, can end up altering the reality of your truth. Lipton argues that this kind of thinking is only confirmed because of the *environment* you grow up in and the beliefs you put on yourself, not so much the DNA itself. In other words, we *think* ourselves into these realities, and more often than not (95 percent of the time) it's not due to conscious thinking. The *code*, aka your subconscious, makes you believe, through repetitive thinking, that this is the truth.

Luckily, we *can* change. We can question our truths. All it takes is becoming aware they are there and then finding the courage to challenge them. Just remember that since the code is set in place to make things easier for you (read boring and safe), retruthing will never come without resistance. However, once you master the art of retruthing, you will not just get better at creating new realities for yourself (and for the world), but learn to love the endless possibilities of becoming someone new.

Retruthing in Practice

As I mentioned earlier, the first time I challenged one of my truths in a conscious way was when I decided to stop eating animals.[8] When I read *Eating Animals* by Jonathan Safran Foyer the summer I was twenty-two, my worldview flipped forever. I would never see myself or my food habits the same way again.

Coming from a family with no vegans or vegetarians, it was foreign for me to think that eating meat and dairy products would be bad. I was raised by good parents who loved the outdoors and taught me to treat animals with respect. My dad was a bird nerd, and I don't know how many hours we spent by the side of the road, binoculars glued to his face, while we waited for him to be done watching whatever bird he had spotted so we could keep driving.

We were also quite the adventurers. We went camping at least once a year, and my parents invested in bike trip gear for the whole family—Mom, Dad, my brother, and me. With our equipment strapped to our bikes, we spent days biking from one campsite to the next. We powered through rainstorms and wind and had many celebratory ice creams that tasted better than any ice creams before. In other words, I grew up a pretty tough child. I wouldn't at all say life was hard, but I learned a thing or two about grit and commitment. I was also brought up with immense respect for the natural world, which is why I was so surprised when I discovered the truth about eating animals and how it did not match how I wanted to live.

---

[8] Although I'm not asking anyone to go vegan, I want to make the point that reducing meat and dairy products as much as we can is extremely important in our work against climate change. Cattle raising, for example, is the main reason for Amazon deforestation, and the factory farming industry forces cows through inhumane cycles of birth and separation. In order to produce milk, a cow must give birth to new calves over and over again, children they never get to spend time with since they're separated from their mothers in the first few days after birth. Imagine the heartbreak?

Similarly, animals like pigs (which are smarter than dogs and four-year-old human children), chickens, and others we eat suffer terrible environments before finally reaching the slaughterhouse where they end up as food on our plates. I think it's up to our humanity to reflect on this system and decide to choose again—this is not the world we want to continue living in moving forward.

I understood it was time for an upgrade.

If you grew up eating meat and dairy like me, you will understand when I say cutting them out of my diet was a terrifying thought. It went against everything my body knew to be right—it was the opposite of my identity and truth.

Growing up in Sweden, I was raised on a diet rich in pork, beef, dairy, and fish. It was my truth that to be healthy and strong, you ate meat, fish, and dairy. When I was introduced to the animal cruelty and suffering involved in delivering food to my plate, the ground beneath me started to crumble.

It made me question if all the things I had grown up believing were true. What do I do without even questioning why? Which of my daily habits no longer resonates with me and who I want to be?

The summer I read that book was the first time I hit a huge roadblock in my life's journey. I realized that the script we operate by is incredibly powerful, and if we never question it, we may live entire lives that aren't ours. How do you know if what you do is powered by your own thoughts and beliefs or simply a pattern created by the things you've been programmed to believe?

I did not want to be someone who causes other species to suffer. I did not want to be someone who indulged their own short-term pleasures at the expense of another's pain. I was horrified that this was exactly how I'd lived my life up to that point. I knew I had to dive deeper. If there were more code I lived by and simply wasn't aware of, I needed to know, and I understood the only way to truly empower myself was to question everything I knew—over and over and over again.

Retruthing is not about a moral compass, nor is there an almighty Truth to strive for. Everyone is entitled to their own truths.

It's also okay to challenge the same truth many times over. It does not mean you change the sails with the wind. To question one's own truths, not just once but again and again, takes courage, and it's a skill that holds the promise of actually changing the world.

In *Think Again*, Adam Grant goes deep into the intellect and courage needed to rethink our beliefs and thoughts. He writes, "We don't just hesitate to rethink our answers. We hesitate at the very idea of rethinking."

How can we get better at rethinking? How do we, in fact, become champions of retruthing and change? Well, it can only come with practice, but as promised, in the next chapter I'll share an exercise to get you started.

CHAPTER 26
Finding a New Truth

You already know that the first step to retruthing is rethinking the concept of being wrong, but what comes next? How does one go about identifying and challenging the view and ideas one lives by?

Practicing retruthing is taking a seat in the audience to watch your own play. It can be a bit uncomfortable at first because you'll begin to realize all the things you do that you had no idea you were doing (awkward), but it's also quite fun, and once you're able to watch the play in full, you can step in and start rewriting. Here is how to put this exercise into practice.

Step 1—Identify Your Truths, The "Whats"

Dedicate a week to some comprehensive retruthing where you investigate your own life. Take a seat in the audience and start paying close attention to your thoughts, habits, and reactions. You want to get a good idea of who you are and what you do so you can start the process of understanding what truths might possibly be in play. Keep a notebook close or start a note on your phone and jot down your discoveries.

- What do you fill your days with? For example, what's the first thing you do when you wake up in the morning?
- What do you buy in the grocery store?
- What activities/things make you happy?
- What do you do that *doesn't* make you happy, but you do anyway?
- Which words and actions trigger you? What makes you feel frustrated, angry, happy, or sad?
- How do you see yourself and the world? What do you tell yourself is possible/not possible? In what way might you be impeding your own growth because of old beliefs about yourself or the world?

Remember that 95 percent of all these things are subconscious behaviors, so it can be trickier than you think to spot them. Taking notes throughout the day can be helpful in capturing behaviors.

Step 2—Identify the "Why"

Once you've acknowledged a behavior, emotional reaction, or thought pattern, try to identify *why* you act, react, or think that way. What is the root of the behavior? Does it stem from conscious learning that led you to make better choices or simply from a habit you can't shake? Are you holding onto beliefs that are no longer serving you that might be worthy of an upgrade? Perhaps you read an article about carbs being bad for you and ever since you've been avoiding carbs. What clues can you find in your memories that might tell you why you do a given thing the way you do it?

- Is it something your parents told you as a kid?
- Has society taught you to feel this way?
- Were you hurt and left with deep wounds?

Often, especially with deeply rooted beliefs and habits, our beliefs and actions can go way back in time. Perhaps you can even identify a single incident you know triggered a particular behavior many years ago, like a comment from a friend that hit deep and never left. Although you often have good reasons for acting the way you do, you might be holding yourself back, limiting your own potential simply because a behavior once became part of your code, and you never gave it a second thought. By identifying the why behind your actions, you're one step closer to understanding your truths and also closer to changing them—and yourself.

Try to find these *whys* without judgment. Shame might arise in this exercise. If it does, I want you to recognize it's a good thing because shame is actually the first step in growth. It may be uncomfortable, but without something telling us something doesn't feel right, doesn't resonate with how we want to be and the world we believe in, we'll never know to change. Shame serves as an indicator, so it should be celebrated, not feared. We'll talk about this more later.

Step 3—Here Comes the Retruthing Part

Once you've recognized a truth and why it exists in your script, imagine how your life might look if you changed it. Write down the truth as it exists now; then change it out by replacing a word or two that better resonate with who you want to be. For me, the biggest retruthing has been about change itself. Here's an example of what my old belief about change was and how I chose to retruth it to open up a whole new world of possibilities.

Old truth: Change is hard and scary. I don't like change because it's full of uncertainty and the potential for failure. The fact that the world around me is so uncertain and ever-changing frightens me. How am I supposed to know what to expect?

New truth: Change is so fun and exciting because you never know what awaits you just around the corner. I love change because it empowers me to become someone new. Life is ever-changing, and I want to be ever-changing too. By embracing change and seeking opportunities to grow, I know I'll truly thrive!

You don't have to believe the new truth right away, and chances are you won't. Remember, your ego operates on old data, so changing your code will feel uncomfortable. That is simply because you have no data in your subconscious to support the new belief. As soon as you start *doing*, you'll add new data to your code, and it will get easier and easier every time you do it. This brings us to step number four.

Step 4—Execution

Once you've identified what you do and potentially why and chosen to experiment with a "truth replacement," start practicing this new truth. What could your life look like if you started living by this new truth? I recommend creating a goal and writing it down somewhere you can see it so you don't get hijacked by your subconscious and thrown back into the old loop. Find a way to remind yourself often of who you want to be next.

It can be as simple as "I choose to be more curious in my actions because I believe change can bring me good things." Each time you read it, you're reminded this is how you want to live your life, and you can begin taking small steps to edge yourself closer to that truth. Maybe you say yes to a new opportunity, maybe you join an art class you always wanted to take, or you do something as simple as taking a different route to work. These small changes start to rewrite the script in your mind, and as you build on the script, it will be easier and easier to believe. If you have been telling yourself you are terrible at yoga for years, but you always wanted to be more in tune with your body, simply changing that truth about yourself could be the first step in getting your ass to a lesson. You will get better with each lesson, and you will also be more comfortable in your new truth. You are simply a living example that changing an idea about yourself is possible.

I know choosing change isn't easy. It can often feel quite scary because your subconscious doesn't want you to change. But the more you practice questioning trivial things, the better you will get at going with the natural flow of change, and soon you'll be ready for the big changes as they arrive at your doorstep.

> *"I love being wrong because that means that in that instant,*
> *I learned something new that day."*
> *— Neil deGrasse Tyson*

Opening Doors to New Worlds

Understanding and mastering retruthing is quite wonderful because not only will you understand yourself better, but you will also be much better at understanding others. With more understanding, we make more room for compassion, which means more room for all sorts of change.

Change starts with you. It can be hard to see at first, but the fastest way to change others and the world is by committing to changing yourself. When it comes to the work needed to tackle climate change, it's essential you understand this. If you are who you are because of your truths, what will happen once you dive into your subconscious and start moving things around? Will you have the chance to create a completely new life, maybe even an entirely new world?

I believe the answer is *yes*, and I firmly believe if we understand the power in our subconscious coded truths, we can unlock our potential for redesigning society and, ultimately, changing the world.

Ideas for Things to Challenge and Potentially Retruth

Here are some suggestions for things to start retruthing if they resonate with you. If one or more triggers you in any way, let them be. Some of these might even seem impossible. That's okay. Find one that has you go, "Yes, I can see myself rethinking my relationship to this," and start there. As you start to retruth small, seemingly unimportant things in your life, cool things will happen. *You* will start to shift and your worldview too. Don't expect fireworks or flowers. Transformation is slow and can be hard to recognize at times. But one day you'll wake up and it will hit you—"Wow, I'm a completely new person living a completely different life!" And that, my friend, is cool.

Don't start retruthing views or ideas that make you angry; start where your curiosity flows. And as you begin to shift with the winds of change, you might be surprised when you come back to this list later and find many items no longer trigger you. The person you are today might think it's impossible, but the you of tomorrow is not you; it's someone else. I bet you can't wait to meet that person.

Truths Worth Challenging

- I need meat and dairy to stay healthy and strong.
- I can't live without my car.
- I can't dance.
- I'm a terrible singer.
- I don't eat breakfast.
- I can't live without sugar.
- I'm way too shy to [blank].
- I'm a woman.
- I'm a man.
- Material things bring me happiness.
- I could never wear the same outfit twice (what a fashion sin).
- I need alcohol to feel confident, beautiful, strong.
- I could never hug a tree—how dorky.

- I'm not smart enough to [blank].
- I'm old.
- I am a terrible cook.
- I'm a bad friend.
- I'm a bad parent.
- People are out to get me.
- It's too late to [blank].
- I don't like healthy foods.
- This is simply who I am.

CHAPTER 27
Shame

It felt like a nuclear attack on my internal system. I didn't know I could feel so much shame in such a short time, nor did I know *where* the shame was coming from. It was so uncomfortable I didn't know what to do with myself, and I had no way to escape it. My usual escapes (walks in nature, music, calling a friend) didn't help. If I tried to step out, it would only follow me on my walk, where I had even more time to think and get stuck in the icky feeling. It was clear. The shame was here to stay, and I knew I had to face it. I clearly had something to deal with, and I needed to figure out what it was.

It was early June 2020, and the words "I can't breathe" were seen and heard everywhere you turned. After the brutal footage of the murder of George Floyd, the world was in an uproar. A couple of months into COVID lockdowns, people were stuck at home with way too much time on their hands and more available than ever to receive this message of injustice. With that footage going viral, it was like we had reached a point of, "Oh, hell, no more!" and the #BlackLivesMatter hashtag spread like wildfire across the United States and the world. We were ready for change.

Cancel culture gained a new meaning in those early days of June. As a white person, you could get called out for speaking up too much or not speaking up enough, or looked upon as an ignorant idiot for posting a black square. There were no rules and so many rules all at once, and I kept hearing people in my circles say in despair, "I just don't know what I'm supposed to do."

As someone who uses her social platform to share and inspire, I felt lost in a world I knew very little about. I was called out numerous times for being insensitive. I was asked to please educate myself to understand better. I felt attacked and, at the same time, I knew the attack was probably valid. I *did* feel insensitive and uneducated, and then I felt shame.

Unbelievable shame.

I had always thought I supported black people. I have always believed all people should have the same rights and opportunities, regardless of skin color or background, so obviously, I thought, *I'm not a racist.*

But oh how ignorant I was. I've learned that being a non-racist is not enough, that my silence and inaction are only allowing the systemic, entranced, centuries-old underlying racism to carry on. It wasn't enough not to be racist. I had to actively be anti-racist. But it took shame to get to this point. It took shame to understand that being anti-racist takes work. Not only that but a majority of that work doesn't happen on the streets carrying a sign. It happens within *you* when you're alone. It comes with recognizing something doesn't fit how you want to think about yourself and the world and then finding the courage to turn that rock over and see what's hidden beneath it. It comes with facing that *shame.* What you'll find is hidden truths in your system, sometimes centuries-old data that's never been questioned and has now simply become a part of your code—fears and beliefs that have been passed on through generations and now ended up within you.

It might be uncomfortable to discover unknown parts of you; it's like discovering a dirty worm in your belly, and once you know it's there, it makes you feel sick to your stomach. You want it to go away, but you can't get rid of it. Anywhere you turn, the feeling is still there, and it can become unbearable to deal with.

When I was called out on my ignorant whiteness, I felt like the worm had grown into an adult anaconda. It slowly became impossible to live with myself. I couldn't work, I couldn't relax, and at times, I could barely eat. I called my white friends for solidarity. I even called my black business partner for understanding and support. Nothing helped until I realized the shame I felt, the worm that had grown into a snake and refused to go away, was actually nothing but a beautiful gift. It was a gift that (if I let it) could transform me. Because now that I had recognized it, I had the chance to finally set the worm free, and with that, give myself room to *grow.*

If I didn't feel shame, it would mean I didn't care. I recognized there are a lot of ignorant people out there who don't give a shit about what they do to the world, and they surely do not feel shame when being called out. They may feel angry and annoyed. They certainly feel they've been unfairly attacked. But they don't feel ashamed. Since I did, in fact, feel a great deal of shame, that meant I *cared*. Not only that, but I wanted to change.

Suddenly, everything appeared in a new light. I didn't want to hide from myself anymore. I wanted to go deep and unlock this hidden gift. I began treating the snake in my belly like a travel companion with great wisdom rather than a rascal to get rid of. I wanted to know what it knew about me that I didn't know.

This insight led to deep transformational work on the gooey insides of my soft and vulnerable ego. I read the books others recommended, like *Me and White Supremacy* by Laila F. Saad, and continued to stir up shame as I started each new chapter and learned yet another thing about myself and the world. Only this time, I didn't run for the hills to get away from it. I looked the shame right in the eye and asked, "What can I learn from you?"

My anti-racism work is ongoing, and I will probably be a student for life. I embrace the work, and I'm proud of myself for wanting to change. I also use this story to remind myself how empowering shame can be. Next time it comes knocking on my door, I want to be polite enough to ask it to come in and share what it has to say. After all, shame is a force to be reckoned with, a companion you want to share a cup of tea and have long chats with from time to time. Don't shut the door in its face. Take a deep breath and soften your ego. Then listen wisely and see shame as the gift it is—a gift of growth and change.

Turn Your Shame into Growth in Seven Steps

1. Acknowledge feeling shame; simply realize it's there. Understand shame is a good thing because it means you care and have come to a new realization about yourself and/or the world.
2. Replace shame with acceptance. Shame in ignorance is disempowering; shame in *acceptance* empowers you to change. With acceptance, you allow yourself to heal.
3. Recognize you want to change something about yourself and identify what it is. Ask yourself, "How can I be/do/feel/think differently about this? How have my actions (conscious or

subconscious) been out of alignment with the person I want to be today?"
4. Think globally, act locally. For big matters like social justice and climate change, it's easy to get frustrated and look to other people to do the right thing. We ask for systemic change and for our leaders to take action, but we often forget *we are the system*—the voters, the consumers, the people creating the culture around us. We have the power to *be* the change, which also means all change starts with us—with you. It starts with changing your heart, moves on to conversations with family and friends, and grows into bigger conversations that start to shift the community you live in. Think big (a better world) but act small (start with your community), and never underestimate the change you activate from within.
5. Act—just act! The best way to turn shame into something empowering is by taking action. Educate yourself and then act on that knowledge. This is how you *grow*.
6. Call yourself out. Old habits die hard, but you can change if you allow yourself to be wrong. As you fall into old habits and shame arises once more, do the loop over again. Turn shame into empowerment and positive change.
7. Repeat. Visualize the world you want to live in, and then be the person who makes that world possible. Your words and actions matter so much more than you think, so look boldly at yourself, and make it your mission to always be aligned with your truth. Be the change you wish to see and continue to change the world around you.

CHAPTER 28
Normal

How often do you reflect on what's *normal?* When does something become normal and how fast do we get used to something new? This was exactly what Arthur and I contemplated over a cold beer on a beach in Northern California on a Sunday afternoon.

The best thing about our two years living in San Francisco was all the hikes only a short car ride away. On weekends, we would often get up early and drive through the mist on the Golden Gate Bridge to the paradise of trails just across the bay. One of the hikes we particularly liked was next to a small town called Stinson Beach. The glittering blue Pacific on the left while you walk along the ridge was one thing, but what made this hike so magical was after you sweat for three or four hours, you get to reward yourself with a cold beer on the beach.

The first time we stumbled upon the little beach shack hidden among the dunes, we were quite beside ourselves. Fresh guacamole, homemade chips, and beer on tap? We must've found heaven.

The only problem was the beer came in gigantic plastic cups. I understand the reason for a plastic cup on the beach, but that doesn't mean I like it. The reality of our plastic-filled world has given me anxiety for years, and each time I find myself participating in this toxic culture, I feel a knot build in my stomach. As the woman handed over the beer, Arthur smiling from ear to ear, I found myself once again struggling with my deeply held values and accepting what is.

However, I've learned with planning, many moments like these can be saved, and I made sure the next time we went I would be prepared.

"Almost ready, one second!" I yelled from the kitchen to Arthur who was standing in the hallway ready to go. Proud of myself for remembering the beach shack, I wanted to find something perfect. The container had to be big enough to hold a beer (we don't want to lose value) and shaped so you could easily drink from it. Hmm, what could we…wait, this is it. I reached for an old marinara sauce jar we'd saved and realized it was going to be perfect. Content with what I found, I packed it in my bag, and we headed out.

After our hike, we drove the short distance to the beach and parked in the gravel lot. Sweaty, tired, and high on endorphins, we changed out of our hiking boots as our mouths started to water in anticipation. Barefoot and free, I reached for the bag with the jar and told Arthur about my plan to try to use it for ordering a beer. Knowing not to question my zero-waste attempts, he agreed we should try and hope they would accept it.

Luckily, they did. Not only did they accept the jar, they got so excited about our thoughtful efforts that they even gave us a discount. Since the jar was quite big, Arthur smugly whispered he thought we lucked out on the volume as well. This was a win-win in so many ways.

Slightly smug and highly content with our smart move, we laid down in the sun and sipped away on our well-deserved reward. It honestly felt like life couldn't get much better. We didn't say anything; we just watched the waves hit the shore as we took turns sipping on the beer.

Then Arthur said something I'll never forget. "Don't you think it's interesting," he said, "that all these people here, all these adults with kids and families, probably consider themselves pretty conscious and thoughtful citizens. Yet here they are drinking beverages from plastic cups they will soon throw away."

"In fact," he continued, "it's interesting to think about what we consider normal and not normal, because really, bringing your own cups and utensils you take home and wash should be the normal thing to do. Using something *once* and then throwing it away should be looked upon as weird. I know it isn't in our world today. I know we've created a system where single-use items serve a meaningful purpose, but realistically, it should be weird. When you look at it that way, we're the normal ones, and they're not."

I didn't answer right away. The moment was too precious not to capture in its fullest, and I wanted to soak in every bit of it. Did my husband—*my man*—just say caring for the earth should be the normal thing to do? I'm not sure he's ever been sexier than at that very moment, and I silently renewed our marriage vows right then and there.

"You're so right," I said finally. "We're so used to disposable things we think it's the sensible thing to do. Think about it. Being a parent basically means constantly disposing of used, dirty things, and you're told that's what good parents do. They throw dirty things out and choose lifestyle hacks that allow them to be more present with their children, all while polluting their children's futures."

Arthur took another sip of the beer and screwed the lid back on. "And look, this jar is spill-proof too. How smart is that?" We looked at each other and smiled. *Yeah, we're good!*

We weren't trying to be pretentious. It wasn't like we thought we were better than anyone else for bringing that jar (okay, maybe a little), but the epiphany it brought us was so interesting it stayed with me to this day. What is actually normal?

Think about the society we've built for ourselves. It is a society where convenience and disposable items, especially plastics, have become the norm. It is a society where what used to be common sense, like mending and reusing, is now rare. It wasn't that long ago when everyone cleaned their milk jars and left them on the stairs to be picked up and replaced, or when you owned one coat for the majority of your adult life and kept mending it to keep for years. I'm sure many years ago when someone first brought plastic cups to the beach, people around them saw that and thought, "Oh, that seems cool, and how convenient—maybe I can get some of those too."

What started as a novel trend soon became mainstream, a new norm. Plastic was so smart, simple, and cool. What we didn't know then, but we know now, is how rapidly plastic is filling our oceans. We also didn't know that these items of convenience would shed tiny little microplastics that are now seeping into our ground and finding their way into our water and food. We didn't realize these single use, disposable items, used for less than a day, would take ten, 500, or even 1,000 years to decompose, or the threat they still represented when they did break down.

Did you know that the average person eats a credit card worth of microplastic every week? Even scarier than that, did you know these tiny plastic particles are now found in human organs, the lungs, and even blood?[9]

These facts are scary and quite hard to grasp but sadly true. It is also true we are dumping the equivalent of one truckload of plastic into the oceans every minute—*every minute*—and that truckload is interrupting ocean ecosystems and finding its way into beautiful creatures' bellies. Currently, we're looking at more plastic than fish in the ocean by 2050. If you've been to a beach recently, you may have wondered what the colored line on the shoreline was. It's microplastics, washed up on the shore with every wave, quickly changing our beautiful beaches. It's sad to think about, and what's even crazier to grasp is that 50 percent of all plastic produced is for single-use purposes. A material that is so sturdy it will outlive any human on Earth is created to be used *once* and then thrown away! Can you think of anything more absurd?

But it's not absurd, is it? When we go to the grocery store, we expect to find everything from tofu to pasta packed in plastic, and we reach for a plastic fork with our lunch salad without a thought. Disposable plastic is simply part of our culture, part of what's *normal*, and so who are we to question why it's there?

The truth is there are a lot of messed-up things about our world today, and it can be quite uncomfortable to wake up to them. Humans are a powerful species, and unfortunately, so far we've used most of that power badly. And as our population grows exponentially, so will the magnitude of our influence on the world around us. We are many, and we must recognize that no matter how small our individual actions might seem, they make a difference. For good or bad, they add up.

> *"It's only one straw!"*
> — *Eight Billion People*

We are social creatures. That means we're constantly looking to our surroundings—consciously or subconsciously—to see what's acceptable, normal, and cool. We want to do what others are doing because it means we can speak to the crowd and fit in. Cultural norms are built on this

---

[9] "Microplastics found in human blood for first time." *The Guardian*. March 24, 2022. https://www.theguardian.com/environment/2022/mar/24/microplastics-found-in-human-blood-for-first-time

system, and before we know it, we will seamlessly conform, rarely questioning the actions we take in our daily lives. As decades go by, norms and culture are passed on to the next generation—that is how generational amnesia continues. We simply don't know any different.

How about we actively choose, right now, to wake up from this generational slumber? Maybe we recognize that the norms we're so used to were ludicrous to begin with and don't serve either ourselves or the Earth. Maybe it's time to really begin looking at the lives we live and why we live them the way we do. It's time to find the courage to create a whole new normal. It will be strange and uncomfortable at first, but as soon as a new norm is created, culture starts to shift, and new systems form to support it.

We usually think something is crazy until it's accepted by the crowd. But instead of waiting for that to happen, be the cool weirdo who says, "Hey, I'll give it a try." Claim your influence and see where it can take you—go out there and change the world.

It only takes 3.5 percent of the population to start a revolution,[10] which means we don't need to change everyone; we just need to awaken enough people to start a new norm. Remember that when you hesitate. Remember you have the power to influence and can activate that power at any time with your actions. (More on this in Part Four: Choosing Empowerment.) You can change the direction of the crowd, and in doing so, take heroic steps in creating a different, smarter, and even better world.

---

10 Robson, David. "The '3.5% rule': How a small minority can change the world." BBC. May 13, 2019. https://www.bbc.com/future/article/20190513-it-only-takes-35-of-people-to-change-the-world

CHAPTER 29
Four Teenagers and a Car

The first rays of sun have the magical power to make everything look beautiful. Even an old gas station with some rusty trucks and a car on blocks are seen in a new light. Something about the colors of the morning sky never gets old, never disappoints. It's the promise of a new day once again.

As Hannah waits for the battery tracker on the electric charging station to go up,[11] she leans against the car and recalls a game she used to play when she was younger. It was the game of magic guessing, the anticipation of anything extraordinary to come. Because, truthfully, you never know, do you?

It's a new day and who knows what this new day will hold? And this place? It might not seem like much on the surface, but maybe this is the gas station where someone decided to finally quit their tedious job, marry their true love, or go back to school. Or perhaps later today, two old friends will pull in at the same time and realize how much they've missed each other. Or maybe none of that will happen, but the thing is, we don't know, and that's the whole beauty of the game.

Hannah stopped playing the game as she got older, thinking it was better to be realistic and focus on the now. But she realized she was onto something back then because the truth is there's actually no way of knowing what events will unfold or what they will lead to. And if we remain open-minded, we don't miss those opportunities when they actually come.

Hannah laughs as she recalls another memory from her childhood. Almost every morning, Dad would walk over to the breakfast table where she and her little brother were deep in their cereal bowls, and say, "Kids, today is a special day. Do you know what day it is?"

---

[11] I choose to make this an electric car. The time when this story takes place is somewhat unknown, but it's a time when even young folks can afford cool, electric cars!

"Mmm, Dad, we know...."

Hannah smiles at the memory. How crazy it would drive them, and how Dad knew that but didn't care. He would always deliver the answer. "It's a day that has *never* been before. Isn't it amazing?"

Hannah always thought this statement was half stupid but also true. And although it drove her nuts (increasingly so as she grew older), that saying stuck with her. When you think about it, what does it mean that every single day, no matter how mundane or boring it might seem at first, is precisely as unique and special as any other day? Sure, some days have more to them, like Christmas, birthdays, or the final day of school before summer. On those days you can always expect exciting things because that's what they're all about—excitement of some sort. However, if you start to pay attention, you realize the most extraordinary events can happen any time of the year or any day of the week. In fact, the most unexpected and exciting things tend to unfold during the less celebrated days.

Hannah's dad was right. They all ought to be counted, and one should be careful not to rule out magic before magic has even had a chance to begin.

Hannah's daydreaming is interrupted as Bella rolls down the back window in their jam-packed car and says, "Come on, girl. Fill it up. The roads are waiting."

Hannah sighs. "I'm on it, Bells. The meter is ticking. Relax."

"Just wanna get going, that's all. Not a second to waste, baby!"

Hannah shakes her head at her best friend's childish excitement. Bella is always so eager to get going. They had been planning the trip for months, all four of them meticulously calculating time, miles between recharging, and desired destinations. How far could they drive each day and still enjoy the drive? What cute towns should they pop into on the way? And what breathtaking scenery said to be hiding nearby should they be sure not to miss?

It was incredible how much information there was as long as you typed in the right words. Yelp reviews, blog articles, IG tags, and YouTube videos—the worldwide web was like an encyclopedia of knowledge and exploration. Tales of previous adventurers who wanted to share their magic and ensure someone else got to experience the same thing. You can't help but wonder, what did people do in the old days?

How did they ever figure anything out?

It had started on the couch one Friday night over some microwave popcorn and a bag of salt and vinegar chips. The TV was on in the background, but no one was paying much attention. TikTok and Instagram, per usual, were stealing the show. It was Jasmine who unexpectedly spoke the words that would change it all: "We should all go on a road trip!"

They peeked up from their screens but didn't answer. Suspicious and a bit wary of Jasmine's often unrealistic and poorly thought-out ideas, the other three silently hesitated. Mike spoke first as he reached for the chips on the table.

"Yeah?"

It wasn't much of a response, but Jasmine wasn't known for having calculated her suggestions. Maybe she had just found an inspiring account on Instagram, a ridiculously handsome couple with a cute dog who made their living traveling the world, and now she wanted to do the same thing. Only this group wasn't a cute couple with a dog—they were four friends in a rural city with sad, low-paying jobs.

However, Jasmine wouldn't have their dim response. They were all surprised when she stood up in front of them, eyes sparkling, and said, "No. I'm serious, guys. We should go on a road trip. Next summer. Let's do it. Let's pack the car and go! We can use mine. It should be big enough for our stuff, and we can take turns driving."

"And where would we go?" Bella asked, grabbing the bag from Mike to claim a few more chips before they were gone.

"Anywhere, really. The destination isn't the point. It's about the road, about being as free as those people in the movies always seem to be. You know, when they roll down the windows and feel the wind in their hair and life seems so exciting and simple."

Jasmine was pacing back and forth as her vision got clearer.

Hannah had to admit she was starting to get excited too. "And where would we sleep? In motels?" she asked.

Jasmine stopped pacing and gave Hannah her full attention, thrilled someone was finally giving her a proper response.

"Why not? Maybe we can research some cheap ones in advance and map it out?"

"Or we bring tents and just pop them up whenever we're tired of driving."

They all looked at Mike, everyone thinking the same thing. *Did he just suggest camping? Mike?*

"You do know that means sleeping outside, right?" Hannah teased him. "Like, that place you find yourself when you leave the house."

"Very funny! Of course, I know what sleeping in a tent means. I used to do it all the time with my dad when I was little. He would take me fishing by my grandma and grandpa's place in the summer."

"Really? And how did this little fisher boy turn out to be…well… you?"

It was Jasmine's turn to tease Mike now. It was a thing among them. They all teased Mike, and they all knew he liked it, even if he never admitted it.

Mike shrugged his shoulders and pulled the chips back from Bella, finding only crumbs.

"What can I say. A man changes. That doesn't mean he can't change back."

Mike leaned back and poured whatever was left in the bag into his wide-open mouth. It was a bit more than he anticipated, and salt and chip crumbs showered his chest.

"No disgusting behavior like that on the road, though," Jasmine said. "It's my car, so I get to make the rules, and I don't want critters breaking in at night because we're slobs and can't keep it clean. Besides, there might be bears, so we need to be careful not to attract any wild animals. I've seen that on TV."

"So, wait—you want to do this? You want to pack the four of us in a car with a couple of tents and drive off into the unknown?"

Bella was mocking Jasmine, but she was unable to entirely hide the excitement in her voice. Bella knew it. They all knew it. Of course, they had to go. The hunger for adventure had been ignited inside all four of them.

"Why not?" Jasmine responded. "Right?"

They silently looked at each other.

"Why not?" Hannah said.

Then Bella filled her lungs and screamed, "WHY NOT?"

A chorus of *Why nots* followed, and, like a canyon in the desert, the four friends echoed it back and forth, laughing and howling like wild animals in the night.

An angry knock from the other side of the wall told them they had infuriated Mrs. Stevens, Mike's sweet but very impatient neighbor, an old woman with gray hair and a cat that kept escaping when Mrs. Stevens took out the trash.

"Sorry, Mrs. Stevens!" they all yelled at once, but it only made them laugh even harder. They couldn't hold it back. A wave of something new had come over them. Like an electric bolt from a clear sky, a deeper sense of knowing had flushed through their bodies. They were lit up, vibrating on the same frequency, as they realized the exact same thing—life was more than this. Life had much more to explore than what they could discover on their social media feeds or TV. All of a sudden it became clear they were missing out, not only on places to see but on the true experience of life. A burning desire to reach beyond their safe little worlds had been planted within them. They had to break free and explore what was out there on the other side.

Where they went didn't matter. The destination wasn't the point. The point was being on the road, embarking on a journey with no idea how it would turn out—and quite frankly, that it didn't matter. The goal didn't matter because it was the journey itself that promised all the cool stuff. All the miles on open roads where they would pump up the music while the wind filled the car, where they would stop in unexpected towns and make friends with unexpected people, learning important things from unexpected lessons.

Hannah's words were unnecessary, but she spoke them anyway. Someone had to say it.

"Let's do it. Let's hit the road!"

Five months, two new tents, and hours of dreaming later, the day had come. Hannah closes her eyes to the morning sun as the birds on the roof above sing in the new day. It is hard to believe it is finally time, but it is. It is time for their journey to begin.

Hannah watches the numbers tick on the screen beside her as the car recharges. And as it does, she charges with it, feeling higher on life as the seconds go by. The air is still brisk and raises goosebumps on her bare arms, but she doesn't mind. They match the tickling rush of excitement she's feeling inside.

Hannah tries to grasp what exactly she's feeling right now. Is it the feeling of new beginnings, that frightening mix of not knowing what to expect while at the same time dying to find out? Or is it the feeling of having landed, not just in the moment, but in the magic of life itself? She can't quite put her finger on it, but it's the strange sensation of having found her way back, after years of being lost, to feeling like she's come home. In this moment of uncertainty, with nothing to hold on to but a map and a jam-packed car, she feels a sense of ease she hasn't felt since she was a little girl. She understands that whatever happens, they'll figure it out. And that as long as they keep driving, they will reach new destinations and discover new worlds.

And that's it—that is precisely it. This feeling of calmness, she now realizes, is all about that. It's about knowing she's not stuck, that she's done missing out on life. Hannah feels like she's about to *grow*. She had no idea how confined she'd felt all these years. She didn't realize how her own perception of life held her back. But it was becoming so clear now as she stands here in this magical morning light. Like a long-caged animal, she's setting herself free. She curses her sensitivity as a tear rolls from her eye, but she can't help it—it's the best damn feeling she's felt in a very long time.

Hannah also suddenly realizes, as she sees the circus going on inside the car, that today is actually not the start of their dream journey. That this is not the morning it all begins. It began that Friday night five months ago on Mike's couch when the idea was born. The planning, the fighting, the hours of trying to map out a route was all part of it. All the afternoons at the local coffee shop planning and the late-night texts sharing links to national parks, blog posts, and weird animals to look out for and avoid, were just as much a part of this crazy trip as the road ahead would be. And as far as they know, this journey might never end. It will probably fill them with memories, lessons, and experiences that will follow them for a lifetime. There is no start or end to a journey, after all, just like there's no start or end to the magic of life. It's all just a feeling of having landed, a constant hunger to find more to explore.

She takes a deep breath of the crisp morning air and stretches her arms above her head. With a few head rolls, she cracks her neck and lets her body wake up. *What a day!* The charging station buzzes, telling her the charge is complete. It's time to embark.

Jasmine, excited, slams the door shut, and lets her car come to life. She turns up the volume as Queen's "Somebody to Love" comes on.

Hannah slides herself back in the seat.

"All right, guys, let's go!"

"Whoopee! Let's go!"

Mike sings along with Freddy's high-pitched tones.

"Oh, shut up, Michael!" the others say in unison.

CHAPTER 30
Road Tripping

Have you ever been on a road trip or at least dreamt of it? You and your favorite people and the open road, with nothing to do for days but watch the winding road ahead and explore new towns. With the excitement of not knowing what comes next, the freedom of having no responsibility or obligations but to keep driving.

Going on a road trip is unlike any other trip you'll ever take. It's not as luxurious as getting on a jet and landing on the beach tomorrow, nor does it rejuvenate your cells like a backpacking trip in the mountains, but it comes with its own kind of magic.

A successful road trip (there are awful ones too, I'm sure) is based on a few things. For starters, you must have a destination in mind, a place you're headed to, but you shouldn't make the whole trip about where you want to be. When you go on a road trip, what you experience on the open road is just as important, if not more so, than the destination itself.

Second, you have to travel with an open spirit. The roads may present you with all sorts of challenges you couldn't possibly predict, and when they do, how you handle the situation will determine everything. You may still make it to point B, but whether you enjoy the ride is up to you and comes down to open-mindedness and attitude.

You can get stuck in remote towns with a dead battery (been there) or drive for hours without a rest stop in sight. In those cases, it's easy to feel like the Universe is conspiring against you and you're dumb for having taken off in the first place. But that's never the case. Road trips come with unknown predicaments, and I'm beginning to believe someone is planting them there because they are *meant* to be there. It's not supposed to be a smooth ride. You will be thrown off-route from time to time because that is how you *grow*.

Think about all the Hollywood movies based on the plot of two or three people—friends or strangers—sharing a car for a day or two. Either a flight got canceled and there was only one rental car left to share, or a group of friends needed to travel to another town for some important business. The movie is never about what happens when they get there (although that's the only thing the characters have on their minds)—it's about what happens while they're on the road. It's about the struggles, the fights, the loud singing to old music, and the start of new friendships while sharing stories by the campfire at night. Movies like these exist because they're interesting. Road trips make people grow, they make them change, and that is why we have fantasied about them since before the car was invented. It's about seeking a journey, but in doing so, finding a part of yourself.

> *"Maybe it's not about the dream. Maybe it's about who we become while we're chasing it."*
> — *Alexander den Heijer*

I'm not selling you hard on road trips because I want everyone to jump into a gas-guzzling car and pollute the Earth for hours on end while seeking your soul's purpose. However, if you do feel called to go on one (I have many times), there are many ways to travel sustainably these days. You can get a monthly train card and explore Europe or rent an electric car. However, I'm talking about road trips because working for a better world is very much like a road trip, and you should do your best to treat it as one. Let me explain!

How Climate Work Is Very Much Like a Road Trip
One—Unknown Destination

The first thing to get clear is we don't quite know where we're headed. A better future, yes, but we actually have no idea what that future will look like. That means we must keep our desired destination in mind while being courageous enough to take off without knowing exactly what we're headed toward. Dreams, visions, and tales of better times can spark the desire to go, but it's not like a charter trip where there's a beach and perhaps an all-inclusive hotel waiting for us at the other end. We don't know what the future looks like. Nothing is promised, but the pure longing to search for new worlds should be enough for us to go.

Two—A Marathon, Not a Sprint

Just like any one-way shot through the American West will have you contemplating everything from your relationship with your mother to your identity and purpose here on Earth, you can be sure this trip will also bring a lot of deep thoughts and reflection. Our journey toward a different tomorrow, our work to cocreate a better world, will only be successful if we reflect on who we are. The world will only change if we change. Broken systems will only be dismantled and rebuilt if we go deep within and rearrange the building blocks of our subconscious.

That is why it is so important we think of this work as a marathon, not a sprint, a long road trip from coast to coast, not a private jet to a remote island. Although we have a goal in mind, an idea of a destination we are teased with from time to time, it's all about what happens on the open road. Who will *we* become while we travel? Will we allow ourselves to go deep, reflect, and grow? That is what the work of our better world journey is all about.

Three—Stop to Take Breaks

Similar to a road trip, you will have to take breaks. If you run out of electricity (or whatever fuel you're using), the car will stop driving and you might break down in the most inconvenient places. If your belly gets too growly and you have to pee so bad you can't think, accidents may happen (accidents of any kind). Any well-rounded traveler knows to stay happy on the road, you need to be smart and rest from time to time. You need fuel and food. Preferably you also want to seek out fun stops along the way where you can go for a walk, visit a waterfall, or get a beer in a quirky town. Those stops are what make the trip worth it. Without them, it's just a tedious drive from point A to point B.

Working to cocreate a better world is the same in so many ways. Success will not happen overnight. Our fight for a climate-just future is not going to end tomorrow. Most likely, to most of us alive today, it will be the work of a lifetime. New audacious goals must be set for each decade, and we need to keep digging up new courage to move forward. Each year, new change will be upon us, and we must be resilient enough to look that change straight in the eye and say, "I'm ready. Let's go."

They call this decade (2020 to 2030) the decisive decade. How we act now will decide the future we live in. For many of us, that means life and death, village or no village, future or no future. It is that serious. But the next decade (2030 to 2040) will be a decisive decade too. It's not like we will reach the finish line and can finally exhale and relax. Once we reach the top of that mountain, another, perhaps even bigger mountain will be on the horizon. We will need to keep going.

If we are to sustain our work *and* ourselves throughout this lifelong challenge, it is essential—and I emphasize *essential*—we learn how to take care of ourselves.

Your body is your vehicle, and your soul is your driver. They are equally important. You need to take care of your car (body), so it doesn't break down in the middle of the journey. If you're unlucky, you'll be stranded in the desert with the scorching hot sun and no gas or water. We need you to be as strong as you can possibly be, so don't underestimate the importance of taking care of that vessel of yours. Don't let it fade because you were too lazy to go in for an oil change.

If your body is your vehicle and needs physical activity and good nutrition, your soul and spirit need love and attention to drive the car forward. You need healthy, delicious foods, good conversations, friends, fresh air, and a full night's sleep. If you deprive yourself of any of that, you might be able to keep driving, but your trip won't be any fun, and you won't be as good at navigating changes and challenges as they appear along the way.

Sometimes there will be traffic jams. Sometimes a road will be closed, and you'll have no choice but to add hours to your trip as Google reroutes you. Other times, you will have so much fun with your friends in the car that you will barely notice time passing at all.

The better you are at taking care of *yourself*—the driver, soul, spirit—on this journey, the better equipped you'll be to continue it. You will find new solutions when roads are closed. You will bring a light spirit and critical thinking to tricky situations and find new ways. When you get tired, you will ask a friend to take over the wheel so you can take a nap, happily knowing you're still en route while you rest your eyes for a while.

I'm talking in metaphors, but it couldn't feel more applicable to the work we're here to do. If we don't take care of ourselves—soul, body, and spirit—it is going to be very hard to move forward. We will hit walls, get drained, and lose ourselves to despair. And as we do, it will be very hard to stay curious and solution-driven, undermining our ability to find new ways.

Four—Friends

Do not forget this part—community and friends are necessary. If we don't surround ourselves with community, if we don't have friends in the car who can share this journey with us, it will get lonely and exhausting. If you are the only one driving, the whole trip relies on you. But if you have friends who can take turns, it's okay to take a nap from time to time, knowing you're still moving forward.

Our climate work is very much the same. Some days you are just not that motivated. Sometimes you think to yourself, *I am in no position to change the world today*, and that's absolutely okay. When those days arrive, take a well-earned break and rest in peace (on the couch, not in a grave) knowing someone else is covering your leg of the journey today. Millions of people in the world are fighting for a just and sustainable future, so know the journey will go on even if you take a break. But it can be hard to remember that if you're alone in your car, so surround yourself with a community of like-minded people. Find people who can pep you up on the down days and sing along in the car when things are fun and exciting. We're doing this together. We're *cocreating* a better world, so never feel like you have to go it alone.

Five—Have Fun

The last thing I'll say, and the most important distinction about a road trip compared to any other trip, is that the fun starts right away. You don't wait until you land at the resort, kick off your shoes, and run to the beach to feel like your trip has begun—the fun starts right now. The second you pull out of your driveway and head for the nearest highway, your journey begins. The thrill starts to spread through your body as you navigate familiar roads and realize you have no idea what awaits. Maybe you put on some music and tear open a bag of chips, all while your friend undoubtedly complains you haven't even left town yet and should save the snacks for later—to which you just turn up the music and roll down the windows because *who cares—you're on your way!*

Our work for a better, just, and sustainable future starts right here, right now, and the fun has to start right now too. Don't wait for the day we seem to have figured it all out because that day will never come. We are figuring things out as we speak and will continue to figure things out, but that feeling of touch down when the flight hits the runway is not going to come. This is a journey of exploration, of adventure, and of one small success after the other. Slowly but surely, we will make our way to a better tomorrow. And then we will need to move forward into another better tomorrow, yet again.

This is what we're in for—seek, find, seek some more. This is the journey of our lives. Change will continuously be upon us, and when that change happens, we have to ask ourselves, "Do I want to park myself in a garage and hide, or do I want to put on my favorite music and keep driving?"

How much you enjoy the ride is up to you.

This is such an exciting time to be alive because we're in for the journey of a lifetime. *Your* lifetime. This is the kind of shit they write about in the history books. You don't want to miss out on this ride, trust me. You want to be here for all of it. You want to look back at these years and proudly tell your grandkids, yes, you were there when it all happened, and they should know that it was a hell of a ride.

We are the children of change, born into a time when only one thing is certain—*nothing* is certain—and when the only thing we have to count on is our hunger and eagerness to figure it out. Things are about to change in ways we can't even imagine, and there's nothing we can do about it. What we can do, however, is get ready for the road ahead.

Buckle up. It's time to get excited!

CHAPTER 31
Keep Some Room in Your Heart for the Unimaginable

I'm about to share a quote that changed my life forever. Maybe it will change your life too.

When I was a little girl, my headstrong personality wasn't that big of a deal (although I'm sure my parents would argue differently). I resisted simple changes, like putting on my stockings when it got cold (why would you, when you can kick and scream your whole way home?) or not allowing my mother to change out the kitchen curtains for Christmas. Seemingly ridiculous fights, but quite harmless nonetheless. It wasn't until my late teens that I realized this way of operating wasn't quite working anymore. For example, making a case to your ex-boyfriend of why dumping you was a bad idea is, believe it or not, a bad idea. Stubbornly sticking to my masterfully made-up "plan" while missing opportunities that were right in front of my eyes didn't serve me well either. I'm sure as my angels were watching me go through all these struggles, one of them finally had enough and said, "All right, this has got to change!" At least that's what I think happened because someone definitely sent me a message that day!

It happened my second year of college. I was living four blocks away from my parents while attending school in my hometown (something I'd promised myself I wouldn't do, and yet there I was). My life from the outside probably seemed great. School was going well, I had made tons of new friends, I was in good physical shape, and everyone in my family was happy and well. I got top grades while working shifts at the local coffee shop and still managed to attend most parties. If you looked at the "script" I had created for a successful life, everything appeared to be perfect, but I couldn't help feeling as if something were off. Something seemed to be missing and whatever that was appeared to grow bigger by the day. Then the message came.

I was slouching on the couch and scrolling through my Instagram as all young adults do when a quote appeared on my feed. My scrolling thumb came to a stop, and suddenly, I got chills up my arms and could feel my neck hair rising. It was one of those *Twilight Zone* moments when it feels like the entire universe comes to a stop and the room temperature drops multiple degrees. I don't know what it was, but something about the words had me completely captured in time. The quote I read was: Keep some room in your heart for the unimaginable.

If you just read that in passing, go back and read it again. Slowly. Let it really sink in! Because what the F does this actually mean? How do you imagine the unimaginable, and what does it mean to *keep room* in your heart for whatever that is?

But then it hit me! My heart had zero room for anything unimaginable. It was as if I could, for the first time, see myself clearly, and it became so obvious I was missing out. I realized that as I tried to plan everything in the tiniest detail, I had never left room for any magic to appear. All I had ever known was living life by that script—my perfect *plan*. But *what if,* I thought to myself, *there are things out there that could happen in my life, things so extraordinary and unbelievable that even if I tried, I couldn't imagine them right now? What if I am on the road to something grand, a future so bright there is no way my ego can ever plan for it, and all I have to do is get out of my own way?*

I realized I had been living my life in mental captivity, blocking every door to the undiscovered as it tried to come my way. Not being able to see and accept things for what they were, I had postponed the life I was actually meant to live. At first, I felt like a fool, but only for a second, because what followed next was this knowing that everything was about to change.

This quote showed me that so many times we're blocking our own paths because we're only picking up clues from what we already know. We try to make sense of our next steps based on what we're already used to, digging deep into our subconsciousness and looking for data to support our next move. But if we only act on what we already know, how can we ever experience something different? It's like trying to have a new culinary experience while cooking with your old cookbook—the recipes haven't changed, so neither will the food!

For my birthday, I asked my mom and dad to print this very quote on a canvas so I could hang it on my wall. I wanted to see it every day and never forget how I wanted to live my life now that I knew how to. I wanted to stay open-minded and excited, with a curious heart and with room for the unimaginable to come knocking on my door. I was so excited to learn what my life would look like next, and each day became another day of curiosity and exploration!

If I were to go back in time and tell my twenty-year-old self that I would soon live in places like New York and San Francisco, meet the man of my dreams at a bar in Brooklyn, become a fashion model in NYC, and go on to teach the world about climate optimism, I would not believe a single word I said. It would seem way too unreal because it was too unlike the reality I was living in at the time for my old self to comprehend. I'm a very different person today, living a very different life, and the only thing that shifted was my perception of change.

Nature changes, as do you. The river keeps flowing, the seasons keep turning, and your body keeps producing new cells. If you try to resist this change, you will only feel frustration, as if you're swimming against the current and the water keeps pushing you back. That beach you left when you jumped in for a swim seems harder and harder to get back to; the struggle to stay close gets tougher as your body weakens in the swim.

I understand that returning to the same beach, the one where you left your towel and your clothes, is a compelling thought and probably the most reasonable option. But what if, from time to time, you let go and let the river take you where the river wants to go? What if by accepting the flow and releasing yourself to the change that's upon you, you will soon reach a beach even more beautiful than the one you just left? What if this new beach is filled with cool people and one of them turns out to be the partner of your dreams? It would be silly not to explore that option then, wouldn't it?

Sometimes life is that river and it will be really tough to swim back. Despite all the reasons it makes sense to return to the place you just left, roadblocks and struggles keep coming your way. In those times, ask yourself, "What am I fighting here? What am I holding on to so tightly I can't see the change and new opportunities ahead?"

When it comes to climate change and our work in cocreating a better world, it is incredibly important we recognize this. We tend to think the world we know now is the best there will ever be, but maybe the future is better and more beautiful than anything we've ever seen before. It's just too unimaginable for us to see it today.

We must understand it's time to move on. It's like the ex who continues to rip your heart open and pour salt into your wounds. If we keep living in denial refusing to admit it's over, or if we keep panicking and ask ourselves, "How did this happen? Why did this come to be?" we're only running in circles, reopening old wounds. It's great we're starting to understand things are not working. I'm glad riots are spreading across the globe as young and old are asking people in power to wake up. It means we're in the heartbroken rage stage—that we're angry and sad. But when we know this, how do we heal and move on?

We've made some serious mistakes. This relationship we've had with Mother Nature has not been healthy. We've taken advantage of her generosity and messed around like teenagers while she continued to provide, so it shouldn't come as a surprise she's had enough. She's asking us to grow up and change. And if we do not, she will kick us out. It's time we understand that sobering realization.

You already know from previous chapters that we must carefully think about the future we want for ourselves and how to actively start choosing the steps that will take us there. Since the future we want is going to be unlike anything we know today, it is critical to leave some room in our hearts for the unimaginable so we can continue to act on our curiosity and find new, smart, and creative ideas.

I have no idea what our future will look like, but as often as I can, I allow myself the freedom to dream. In the next chapter, I will help you retruth the concept of a dreamer and share some tips for actively becoming a dreamer yourself. Because the truth is we need crazy people in our world today, crazy people who are not afraid to create moonshot missions and believe the unbelievable can be real. To quote the brilliant Charles Eisenstein (speaker, author, and crazy person who seemed to have it all figured out):

We have to create miracles. A miracle is not the intercession of an external divine agency in violation of the laws of physics. A miracle is simply something that is impossible from an old story but possible from within a new one. It is an expansion of what is possible.

What future could we find ourselves in if we kept some room in our hearts for the unimaginable? What life could *you* have if you allowed yourself to dream and believe? Too often we're bound by our current reality, stressed and filled with worry and fear because we don't know what the future will look like. We fear change because we don't know what change will bring. But we know everything we've ever built, invented, or created came from an ability to look beyond what we know to be true now.

Our world today consists of millions of manifestations—big and small—from the minds of people who had the courage to dream of something new. People who were brave enough to think of realities that lingered outside their comfort zones and dared to believe in the unbelievable and imagine the unimaginable found the courage to explore those new worlds.

When it comes to climate change, the only hope we truly have is found in that courage. We must expand our hearts and minds, begin to question everything, and dare to dream. Because maybe change isn't so bad. Maybe change means things could be even better, better in the most unimaginable ways.

Exercise—Start Dreaming

Find a quiet moment and get into a comfortable position, sitting up or laying down, whatever feels best to you—but try not to fall asleep. I find it helpful to put on some peaceful music so I can really get into a different state of mind. Then simply ask yourself, "What future is possible if we only keep some room in our hearts for the unimaginable?"

Let yourself drift away. Maybe you start getting visions, or maybe you don't see anything, but you reach some sort of feeling instead. If so, it's totally fine, because remember, we can't *really* imagine the unimaginable, so a feeling is more than enough. Perhaps this feeling is reason enough to keep believing? Maybe this feeling—be it a feeling of calm, excitement, love, curiosity, or joy—will serve as fuel to keep you going. It may be a reminder that a deeper part of you knows a different world is possible if only you continue to work hard and believe.

If you are getting visions, what are they showing you? Try not to dismiss them too fast because you think they seem silly. Remember you actually have no clue what the future holds, so let your imagination roar and trust whatever your inner eye shows you.

Once you feel ready to come back to the now, slowly ease yourself back into the room you're in. If you feel like it, you can end your session by journaling about what you just thought of, felt, or saw. And if you need help tapping into your imagination, I have created a guided meditation you can access on my website, www.TheClimateOptimist.com.

Remember the future you see is already a reality inside you, and you can tap into it whenever you want. And when you know this, when you're aware that this future lives inside your heart, how can you carry that with you into the day? How will you show up in the world knowing this is what your heart desires? What actions will you take to pull this future closer? What choices will you make to fuel this vision and keep the feeling alive?

Remember anything is possible if we only dare to believe, and the first step to a different kind of world lies inside you, in a place where you find the courage to let your visions be free.

You don't have to know all the answers. But as you visit this future and *feel* it with intention and love, you begin to manifest it. Your life will begin to shift, as will the world around you. You will see things you haven't seen before, and what you believe is possible, true, and meaningful will begin to shift and expand. A seed of change will have been planted inside you. How will you nurture it so it can grow?

Take a final deep breath and tell yourself: I will make sure I always save some room in my heart for the unimaginable.

CHAPTER 32
A New Future

What can we—the children of today—do if we only dare to dream and believe? What future will we find ourselves in? What worlds will we be amazed by and love?

Imagine a future where we've found a way to get rid of fossil-fuel-powered cars. It will be a world without much of the pollution we know today—no air pollution, no noise pollution, no toxic chemicals in rivers and seas. Instead, cities will be filled with different sounds—music, birdsong, people talking and laughing, the sound of life. That reality can be ours, and believe it or not, we *are* edging closer to that future by the day.

I want you to imagine a future where cities are more green than gray. In our efforts to stabilize microclimates in urban areas and bring air quality back to healthy levels, trees and other greenery have been planted throughout. Plants are hanging off walls and balconies, trees are in every corner, and rooftops and parking garages have been made into gardens where you can harvest your produce locally instead of having it trucked from miles away.

It is also a future where we've reconsidered the meaning of life. With boringly on-time public transportation and a new work culture that enables people to work closer to home, combined with an increased emphasis on community and family, we will have more time to spend with those we love. And without the stress of driving, finding parking, or commuting long hours, we can be there for what is actually important—a job we love and more free time to cook our own food, grow our own veggies, play soccer with our kids, and other fun things.

With a circular economy that enables materials and products to be repurposed and brought back to life, very little is wasted, and after years of successful ocean cleanup, the oceans are finally almost free from plastic again. That collected plastic has been repurposed into building materials for housing and roads and is serving a much better purpose than the single-use items that plastic is produced for today.

This world might seem crazy to some, and just over the horizon to others, but the truth is it could or could not come true. Whether or not it will be is for our curious minds to see and for our inspired spirits to work toward, bringing us closer. In Part Six: Climate Optimism, I will share many of the exciting projects already underway and provide as many examples as I can to give you reasons to believe in the possibilities of an even better world. For once you start to believe in the process, you see there is hope, and you can begin to support these seemingly bold efforts. You can help get new bills voted into law or invest in projects you believe in. It doesn't always have to be a lot—sometimes as little as speaking up for change is enough to help fuel the transition.

However, anyone can cheer on what's already working. The difficulty lies in understanding we need change even when it may not seem evident. Take a city for example. Walk around the streets for a while, and look up at the buildings around you. Isn't it fascinating how it all coexists? The plumbing, the electricity, the heating and cooling, the people living individual lives crammed together in massive buildings—and it's all working. Aside from occasional hiccups, life in cities works, and people for the most part seem to be thriving. Yet we need change more than we can possibly imagine. Although we can take pride in what we've already built, we need to bring those blueprints out of storage and think again.

Buildings need to be retrofitted to be more energy efficient. Heating systems and stoves need to stop using gas, and new infrastructure needs to be put in place to easily provide renewable energy to every building. Windows should be replaced or better insulated. Public transportation and roads need to be remapped to transport people more efficiently around the city. And, as mentioned earlier, greenery should play a whole new role as the primary element of building interiors and streets.

Even though cities might look fine and be more efficient than other areas, we have a tremendous need to change them, and we must find the will to recognize this and take action.

With all this change at our feet, it might seem like a long journey ahead before we get to the image described above. But remember, every journey starts with one single step, and also recognize the journey to a more just and sustainable world has actually started. We're already on our way.

CHAPTER 33
The Human Agenda

When I looked out the window that morning, the rain didn't seem so bad. A little trickle, maybe, but not something worth fussing over, so I decided to skip the jacket. It was early fall, and the temperatures were still pleasant enough for a light sweater. I had heard that flood warnings were in effect across Manhattan, but a downpour didn't seem to have started yet, so I decided to play it bold. I could survive a walk in this weather.

However, when you live fifteen floors off the ground, you can get a little…what should I say…detached from reality. When I stepped out onto the street five minutes later, the pavement was glimmering in the streetlight as heavy drops hit the already deep puddles. I knew I wasn't dressed for the task. I was going to get wet—drenched cat kind of wet.

I had a few options: A) I could go back up and change into something more appropriate, B) I could call off my morning walk altogether and wait for a better day, or C) I could say screw it and just go for it anyway.

I didn't feel like taking the elevator all the way back up to grab a jacket, so that wasn't an option. I am not a quitter, so giving up on my walk altogether was also out of the question. Instead, I chose option C. I decided to embrace the rain and just go for it. I would get wet, but so what? I plugged in my music, buckled my fanny pack underneath my two layers to keep my phone dry, and dove out into the rain.

It didn't take long before my sweater got heavy and my ponytail was sticking to the back of my head—it was raining hard. Understanding I wasn't going to be out for too long before I had to turn around, I picked up the pace and turned my walk into a run. I don't run often. (I blame it on bad knees.) But this run was glorious. I felt crazy, wild, and free as a bird, and the streets being close to empty just made me feel even cooler for being out. Sometimes it's fun to be crazy alone, to think you're the only one strange enough to do this.

I was higher on life than I had been in a long time. I kept running, and as I was passing through lower Manhattan's Battery Park, get this, Leon Russell's version of Bob Dylan's "A Hard Rain's A-Gonna Fall" randomly came on my playlist. I don't know about you, but that's what I call magic. The synchronicity made me actually laugh out loud, and suddenly, I felt as if I were in a movie. Since the rain had already drenched the flat NYC pavements, I had to jump puddles as I moved along, making the run feel like a true adventure. Not for a second did I regret having dived out into the rain.

Finishing off by the Hudson River, I paused to snap a selfie with the Statue of Liberty behind me when something quite profound hit me. At that very moment, high on endorphins and drenched to the bone, I realized this was what life was all about. This commitment to just doing it is what it all comes down to. That is our human agenda—we simply have to say *yes* to action and change.

Did you know that we daydream 30 to 50 percent of our days?[12] When we do, we're able to replenish our brain calories and save them for the important decision-making stuff later. Believe it or not, making choices takes a lot of brain calories. So much, in fact, that if we have too many choices to make, the choices themselves start to stress us out and take away from the value of life. I'm sure you've experienced this when choosing a dish off a way-too-crowded menu or having to pick the jeans with the perfect fit. With so many options, how do you know you're making the absolute best choice?

What hit me at that moment in the rain was that sometimes the best thing we can do is stop being so damn meek and just dive right in. We have to stop second-guessing everything and simply take bold action. Once we do, the choice will feel obvious, and everything from there on will come from a place of possibilities and growth.

I decided to go for a run in the rain and experienced magic. As a world, I think we have to dive headlong into *change*. Period. We have to trust what comes next. We have to leave the comfort of our old lives and

---

12 "People spend 'half their waking hours daydreaming'." *BBC News*, November 12, 2010. https://www.bbc.com/news/health-11741350

find ourselves in the curious search for something better. It might seem both silly and scary at first (why leave something we know to be good?), but as soon as we take that plunge, everything will be clearer.

When you're already soaking wet, there's no more deciding whether or not it was a good decision; you just put that smile on your face and run.

> *"We see global warming, not as an inevitability but as an invitation to build, innovate and effect change, a pathway that awakens creativity, compassion, and genius. This is not a liberal agenda, nor is it a conservative one. This is the human agenda."*
> *— Paul Hawken*

CHAPTER 34
We're Going

If you feel we live in a difficult time, if you find it unfair we have to deal with past generations' messes, I get it. At first glance, it might seem we were dumped on this planet in the eleventh hour, and we have to fight a fight that should never have been ours. Just the fact that we're unsure if we have a livable future to grow up in, and for our children and children's children to grow up in, is pretty messed up. It's even more messed up that scientists have been warning us about climate change for decades, yet here we are facing the very disasters they've been talking about for years.

If any of this makes you angry, afraid, nervous, or depressed, know that you're not alone. People around the globe—kids and teenagers in particular—are starting to develop mental distress disorders due to climate change. It's called climate anxiety or eco-anxiety, and we will talk more about it in Part Three. Chances are you're struggling with this too because if you're told your future doesn't look bright, not because you're not paying attention in class, but because generations before you have used our earth like trash, you have a right to feel angry and hopeless.

I feel like that some days too. Sometimes, despite all my efforts to be a climate optimist, I fail, and all I feel is anger and resentment, and I am unable to see any light ahead. But then I hear the birds sing, and I see the trees spring back to life like they do each spring, and I remember Mother Nature has the will to bounce back. She is resilient, and she will continue to fight. As long as we choose to fight alongside her, we can get this right.

The truth is, we're all in this mess together, whether we like it or not. We can choose to see it as the end of time, *or* we can see it for what it *really* is—a time for change. We know what to do to turn this ship around and start setting the course for a better world. We have the ideas, the science, the technology, and the knowledge. We have all we need to start getting this right. The only thing we're missing is the widespread belief that it can be done, paired with our collective willingness to embark on this exciting journey. But if we can begin spreading news of an optimistic future, if we start planting seeds of hope for a better world, maybe we'll see a whole lot more people signing up to help us move forward.

If you think the time we live in is a bad time to be alive, think again. I happen to believe it's the most exciting time ever. It's a time when we all get to come together and create history. Like an exciting road trip, we get to pack our bags and embark on an amazing journey. We are unsure of what's to come, sure, but we are eager to find out.

I've made my decision—I'm going. And if you want to come with me, if you choose to jump in the car and travel these roads by my side, please be my guest. An even better world *is* possible, so what are we waiting for?

PART THREE
AWARENESS AND HEALING

CHAPTER 35
Awareness Crashing In

"Be soft. Do not let the world make you hard. Do not let pain make you hate. Do not let the bitterness steal your sweetness. Take pride that even though the rest of the world may disagree, you still believe it to be a beautiful place."
— Iain Thomas, "The Fur"

We were driving home from date night in the Mission District in San Francisco when it came out of nowhere. Shocked, Arthur looked over at me. "Honey, what's wrong?"

"I don't know what's wrong!" I yelled. "Everything is wrong. This world is wrong. We're wrong. And I am so, so, so mad!"

It was as if a bomb had gone off inside me. I started banging my hands against the dashboard as tears streamed down my face. I couldn't hold them back.

Everything about our night had been great. We were in love, we were happy, and we had family who supported us and our love although they were miles away. There were so many things to be grateful for, and now we were headed home with a cozy night in to look forward to.

But that was before the explosion.

Arthur sat quietly for a minute. He kept driving. Then he asked again, "What do you mean everything is wrong? Are you not feeling okay?"

I didn't know how to answer, and the inability to put words to my feelings only made me angrier. Wasn't it all so obvious? Wasn't it clear the world was fucked up, and we weren't doing anything about it?

As we approached our home, Arthur asked if we should circle the block a few times so I could finish crying. Despite the extra carbon in the atmosphere (which I was way too upset to think about at the time), I said yes, so he kept driving and I kept crying.

When we finally parked a couple of blocks away from our apartment, I had calmed down, but I still wasn't quite myself.

"Do you want to walk home?" Arthur asked kindly.

"No," I answered. I wasn't ready to go home just yet. I felt too lost and upset to let this be unfinished. I felt foreign in my own body, and I didn't know what to do with myself.

"Well, maybe you should call Kerrin?" he suggested. As soon as he said it, I knew it was the right thing to do, maybe the only thing to do. Kerrin was my best friend back in New York and my speed dial for dealing with any kind of emotional baggage. I knew if she saw my random call, she'd pick up. She'd know it was serious.

With a kiss on my forehead, Arthur left me with the keys and told me he'd be waiting with popcorn when I was ready to come up. Alone in the car, I called Kerrin. Luckily, she saw my call and picked up. She couldn't help me understand where these emotions were coming from, but at least she was a friend to talk to, and she was able to ground me enough that I could leave the car and walk home.

• • •

When I first decided to become a climate optimist, I thought the answer to doing so was to seek the good news and ignore the bad. I thought if I could only filter out all the negative and painful information about climate change, I could continue wearing the smile I was given and start to believe in a better world. I wanted to be the sun in all the clouds, for myself and for others, and continue to show up with optimism, action, and faith. If I gave in to the doom and gloom of climate despair, I was afraid I would lose myself back into that world and not be able to continue on my mission.

I had no idea how wrong I was.

What I had to learn the hard way is that there's no such thing as not knowing. Because even if you're trying hard—really hard—to focus only on the positive, a part of you still pays attention to everything else, and your body will remember the pain. While you're out there spreading sunshine, your subconscious picks up all the other valuable information it thinks you ought to pay attention to and stores it for later. And if you never give it any attention, that information will add up and keep adding up.

It took me some years to recognize this, but one day, the outbursts started to become more and more frequent—crying in the shower, tantrums at the salad bar, foul moods over nothing in the car. The "attacks" always came without warning, and I never quite understood what was going on. I seemed to be holding on to something, but I didn't know what it was. Arthur, even more clueless, always tried to help, but he soon learned it was better just to leave it alone. I would come out of it eventually.

That is what having climate anxiety has done to me. I've cried in the shower. I've thrown adult tantrums over plastic forks. I've started fights with my family about things I can't remember. I've been angry for no reason, sad for no reason, and for years, I was haunted by this deep and uncomfortable feeling that something wasn't quite right. I've felt alone in the world as if no one truly hears or understands me, which only strengthened my anger toward the world. I've hated society. I've hated people and their ignorance. I've even hated myself many times. How could we live like this? How could we be so ignorant and selfish when the natural world is desperately crying for our help?

Climate anxiety is real, and it can show up in many ways. My low was probably when I was consciously starving myself because I wanted to feel the pain of the millions of suffering factory farm animals around the world. I thought if I could feel pain, I could somehow connect my pain with theirs, and we would be in it together. This might be the stupidest thing you've heard, but when you're upset, you often act in ways that don't make sense.

CHAPTER 36
Ignorance Is Not Always Bliss

I didn't understand it at the time, but denial was coming back to bite me. Hard. But what I know now, since I'm in my thirties and slightly wiser, is you can't choose not to know. I've learned ignorance is not always bliss, and whatever you close your eyes to will find its way in. In the following chapters, we will explore what it means to be aware, accept our feelings, and channel them into something good. Because believe it or not, we *can* be aware of climate injustice and still be optimistic at the same time. In fact, as I've learned far too well, being aware and grounding yourself emotionally is the only way to truly be a climate optimist.

I will teach you what I've learned over the years, and by applying some of the tools and exercises I'm about to share, I hope you will begin your journey toward becoming an empowered citizen and climate optimist—not free from anxiety altogether but equipped to deal with whatever fear and worry may arise. What works best for you, only you can know. What it really comes down to is committed work and practice. Unfortunately, you can't buy sanity in the store. As with a healthy body, you have to keep doing the work, but the feeling you'll live with is worth it.

Before we dive into the nitty-gritty of climate healing, let's talk a bit about denial. An African proverb I love goes: *You can't awaken those who are pretending to sleep.*

It makes me think of when I was little and how my dad had to carry me in from the car after a late-night drive home. I would do everything in my power to pretend I was asleep so I wouldn't have to walk the few steps into the house on my own. It didn't matter how much they slammed the car door or accidentally bumped my legs against the wall, nothing in the world could wake me.

In Sweden, we call it "fox sleeping," indicating that you're being sly as a fox (although I'm sure most parents can see through the cute act of their children trying to snooze their way out of responsibility). It's not just kids in cars who fox sleep. We all do this on a consciousness level, a lot more than we might know, and with much deeper consequences than a parent's numbing arms.

You can't make someone see what they don't want to see. We have a lot of "not-seeing" going on in this world and for many different reasons. Some people don't want to see because it's easier that way, especially when your business benefits from other people's suffering. Exploitation is, unfortunately, the dirty foundation for many money machines across the globe, with a lot of strategically planted ignorance and denial. When it comes to climate change, it tends to get even worse, not just because companies can make money fueling climate change and, therefore, have a vested interest in spreading denial, but because on purely a consciousness level, the facts can—for any of us—be too hard to take. Looking the other way is simply the easier option.

Katharine Hayhoe, an atmospheric scientist and Christian living in the heart of Texas, might know more about this than anyone else. Having spent years of her career trying to raise climate awareness in communities that mostly believe it's nothing but a hoax, she's dealt with all levels of denial. In my and Robin's conversation with her on the *Hey Change Podcast* (Episode 86), she shared some funny stories about people who made one false argument after another trying to convince her she was wrong. Even though she could point to 4 million scientific studies showing that climate change is real, and she was one of the authors of these studies, they thought they had all the answers.

To say the least, the level of conviction people can fox sleep with is quite fascinating. When we don't want to believe something is real, we travel miles into the absurd to avoid seeing it.

Some people choose to be ignorant because they make a lot of money, not knowing their denial causes a ripple effect. Others who make a lot of money intentionally invest in spreading doubt, confusion, and polarization about climate change so others will remain ignorant. The climate question became highly politicized because people making money on fossil fuels understood the danger of us knowing we were causing it, so instead, they planted doubt and hired people to confuse us about climate change.

When political polling on climate change began, the numbers were pretty even across the board. In fact, Republicans even polled higher when asked if they believed climate change was real, a big difference from recent decades when they tend to fill the climate denial corner pretty well. But then we learned that climate change is much closer than we first thought, and to address it, we have to act here and now. How? By cutting down and weaning off of coal and fossil fuels as fast as we possibly can. Unfortunately, enough people were making a lot of money in those industries that they did not want to see that happen. Maybe they thought their wealth would protect their children and grandchildren. Maybe they didn't care. When greed is your only god, you are capable of great indifference to suffering. And so instead of changing, instead of investing in new, cleaner technology, they hired fake experts to plant doubt.

There's a lot of dangerous fox sleeping in the world, and it's harming people, animals, and the planet. This sort of denial should be brought into the light of day once and for all so we can make healthy and deliberate choices moving forward. It's important to recognize the almost systemic fox sleeping that exists so we can become more empowered in our individual actions and say "no" to what we don't agree with. But it's not just people with agendas who fox sleep; we all do from time to time.

> *"People prefer to learn information they think will make them feel good, and so they seek out good news over bad news. When people suspect that bad news is coming, they may avoid the message— even if this ignorance can hurt them."*
> — *Tali Sharot, Neuroscientist*

Setting evil corporations aside, how is this "not seeing" affecting us? What happens in our bodies when we choose not to know?

You might be surprised to know this, but the attempt at ignorance is actually making you *more* stressed, not less. In *The Influential Mind*, Sharot shares an experiment conducted at UC Berkeley in 1972. Eighty young male volunteers were tested to see what they preferred to listen to during an hour of receiving electrical shocks on their skin. Would they prefer to tune out with classical music, or would they rather listen to a silent station that provided warnings a few seconds before the shock?

It should be added that the information station also provided the opportunity to press a button to stop the shock from happening, yet 25 percent of participants chose to listen to the calming music instead. Maybe the music would help ease their minds and reduce stress? This was also tested, but no, the men listening to the music channel showed higher stress levels than those in control of the situation. Intrigued by their findings, psychologists James Averill and Miriam Rosenn conducted another study, but this time without the option of preventing the shock, only between listening to music or learning when the shock was about to happen. Again, results showed that those who chose to be in the know had less stress than those who were trying to "tune out."

> *"What this experience shows us is that even if you think you will be better off not knowing, putting your head in the sand may end up making you more anxious."*
> *— Tali Sharot*

In other words, denial can hurt you, so if you think it's best to tune out and ignore the pain, think again. It will probably be much better to look the truth in the face and be aware of the circumstances, even when you can't fix them.

If you're absolutely oblivious to what's going on in the world, maybe you can live on your pink clouds and sip lemonade all day and be fine. But if you have at least some sort of foot on the ground, you're going to get burned. It's like standing in a wildfire with your head toward the sky and saying, "Look how beautifully orange the sky is," and then being surprised when things suddenly get very, very hot.

As soon as you accept that ignorance is not always bliss and commit yourself to the land of the knowing, a whole toolbox of empowerment will land at your feet. What I hope will be clear after finishing this section is that *you* are the most powerful vessel of change there is, and by treating yourself and your mental health with love and respect, you can become a guiding light for change moving forward. Learn how to empower and ground yourself and you will be able not just to change your own life, but the world around you as well.

> *"Our sadness, anger, anxiety, and even feelings of hopelessness and depression—when skillfully worked with—can be sane, adaptive responses to climate distress."*
> *— Jack Adam Weber*

CHAPTER 37
Awareness Hurts and That's Okay

One of the most profound things I've learned is that awareness hurts, and that's okay.

I know it might feel like some knowledge is simply too painful to handle, that even if you wanted to know, your heart can't take that kind of pain. And even if it could, what good is it? If you can't do anything to change it, why subject yourself to that despair? Why would you want to think about animals burning in wildfires across the world or children who have to drink dirty water, get sick, and maybe die? Why would you want to know that your future grandkids may be unable to enjoy nature's wonders, the ones you've been so blessed with, simply because our current world is fueling ecocide and ruining their chances?

You take on awareness because your heart *can* take it and because even though awareness hurts sometimes, it's okay. It's okay not to feel great about everything, even as an optimist. The world is far from perfect. I would even go so far as saying it's a fairly painful place to live, especially with today's technology allowing us to gain access to this pain, even if it isn't ours.

Thanks to both media and social media, you can know about suffering and disasters from all corners of the world, leaving your moral compass always challenged. In that regard, awareness can be a real bitch. But it can also serve as the door to all the wonders and beauties of the world. The trick is learning how to keep those gates open without losing yourself and wallowing in despair.

What I understand about awareness is that without it, you're not fully present. You're not quite *here*. Sure, you can consume information quickly, but how are you actually taking it to heart? How much are you actually allowing to register? Similarly, you can live your entire life doing all the things you believe are expected of you, but your lack of awareness of *your own truth, hopes, and desires* leaves you wondering if you've actually lived at all.

That is what awareness does to you—it allows you to *live*. You can live life dipping your toes in those lukewarm waters by the shore, or you can choose to dive in and swim with the dolphins. You can spend your days caring about other people's social media stories, or you can go out in the world and create your own—not stories to share on your feed, but stories just for you, *your* story.

To be the center of your own saga, you need to be fully here. And to be fully here, you need to be aware; aware of your surroundings, your emotions and feelings, how your actions affect others, and how other people affect you. By becoming aware, you dive into the depths of something scary, because suddenly you're not just dipping your toes anymore—you're swimming with the dolphins. And yes, that can be intimidating, but it's also where the magic comes into play.

Awareness is kind of like magic. With those gates flung open, you will start to see and feel the world around you in a whole new light. You will pay attention to all the beautiful things around you, things that allow you to have hope in the world, like the small acts of kindness you *will* begin to see between people. When you're fully aware, you will start to notice how someone at the bus stop lends a hand to help the older person up the stairs or how someone stops to listen to the homeless person playing violin and drops a dollar in the bucket.

Noticing these small moments will begin to transform your life. You will start to understand how much they actually matter, which will make you feel like you matter too. You will begin to see that your actions, no matter how small, matter.

That is what awareness brings—the feeling of connectedness, hope, and that what we do makes a difference. In other words—awareness allows you to see *love*.

However, once you open those gates, everything will gain entry. And the truth is there is also a lot of ugliness in the world. Just as you can now see and feel the love, you will also be able to truthfully see and feel the pain. However, and here's the point I'm trying to hammer home, your heart can take it. Not only can it take it, but in feeling it all, by being brave and saying yes to life and all the messiness that comes with it, you activate what we all need so much right now—healing.

The question you ought to be asking yourself is why those awareness gates closed to begin with. If having them open is the key to truly *living*, why are we so bad at staying aware? There are probably a thousand reasons, but here are two worth paying attention to. First, our focus is literally being stolen,[13] and the devices we keep walking around with make staying aware awfully hard to do. Second, it's hard to *want* to stay aware of the kind of information we're introduced to every day (thanks to those tiny devices).

Let's be real. We consume a ton of potential shit every day. I say "shit" because a lot of what we learn daily is pretty shitty (i.e., doom and gloom) and also shit because half the time we aren't sure if what we learn is even true. How often do we actually have time to fact-check everything we hear and read?

We're keeping our information gates open for business at all times, and no one is at the door checking ID or proof of vaccination. No one is making sure whatever passes through our veil of consciousness actually benefits us—or the world—or if it's even something we believe in. Often, what we let pass through makes us stressed, worried, and even sick, but how often do we take a second to reflect on whether we feel shitty for a good reason?

We live in a super-connected world, and because of it, we have access to a lot of information. Often this information is accessible with just a swipe of a finger. Since happy news doesn't sell (and because, let's face it, we humans are drama queens), what we learn can easily make us feel like the world is going under. Yet we keep consuming negative news, after negative news, after negative news, like it's our civic duty to learn everything there is to know.

What we don't do is allow for the emotional space to process what we're learning. We don't take the time to process. How can we care about children starving in refugee camps or wildfires running amok when our life is filled to the brim as it is? How would we even function if we were to

---

[13] Read *Stolen Focus* by Johann Hari if you want to learn more about how your focus is being stolen.

tap into and *feel* each devastating news story?

If you ever catch yourself feeling overwhelmed, give yourself a break. We are exposed to more information every day than our ancestors were in their lifetimes.

It's easy to get overwhelmed by *one* of these stories, not to mention a trillion of them, so honestly…how do we deal? How have our fragile hearts not crashed yet? The answer is we find ways to numb ourselves. We simply close those awareness gates so we can stay informed, yet not quite *aware* of what's going on. TikTok, Twitter, Instagram, dating apps, Netflix, YouTube, video games…distraction after distraction makes it a little easier to stay dutifully informed, but with clouded hearts.

What we're dealing with here are two unattended gates, and they both need to be staffed with new people. Gate number one is the doorway through which we allow all this crazy information to pass. We believe the more we know, the smarter we come across, and the more accepted we'll be by the people around us. That is why we keep those gates wide open—because we want to be "in the know." We also want to be able to showcase this knowledge to other people, so if someone shares something that seems to be worth knowing about, an Instagram story, for example, we might be triggered to share the same information so you can show you're in the know too.

Don't take my word for it. As I mentioned earlier, a Harvard University study showed when people were given opportunities to share their "pearls of wisdom," their brains' reward systems were activated.[14] In other words—we're addicted to sharing information.

That's how it goes. In, out, in, out, in, out. Information comes and goes with no one at the door checking IDs. I'm not suggesting we shut these doors completely, but I think we should be more careful with what we let through. How much information do you consume daily, and how much of that are you actually allowing yourself to process? We will talk more about this in the next chapter, but begin by accepting that, yes, maybe you might be consuming too much.

If we need ID checkers at our information door, we need *bouncers* at our other door—the door to awareness—because what most of us don't

---
14 Cited in *The Influential Mind* by Tali Sharot.

keep open (and sometimes keep sealed with chain and lock) are the gates in which we allow ourselves to *feel*. Feeling takes awareness. Processing the things we read and learn on an emotional level takes courage and strength. But what I've learned, which has been such a revelation, is if we keep the awareness gates closed for too long, with the information gates wide open, the pressure on those doors builds up. Like an angry army ready to attack, the burden can soon become too much. And then one day, often when we least expect it, those doors come crashing in.

By understanding awareness hurts and, actually, it's okay, you put your party suit on and let those gates spring open. You let every visitor know they're welcome—stress, fear, happiness, joy, pride, anxiety, and more fear—they can all come in. And you let them in because you understand they all come with important information.

By paying attention to what worries you, you can look into the issue and hopefully do something about it. By allowing yourself to truly feel happiness and joy, you'll be much better equipped to deal with whatever comes your way.

With a castle filled with visitors from all walks of life, you can run a kingdom built on equality, justice, empowerment, and trust. Trust in yourself and your capability to process information, paired with the ability to use these forces to act on your thoughts and concerns. Without your awareness, gates open; you're nothing but a passage for information—a "free zone" that stronger powers can take advantage of. But when you learn to have a sensible and steady stream at both doors, you become whole—not just whole, but powerful, trusting, loving, and incredibly strong.

> *"When we compassionately face and let ourselves be transformed by our dark emotions, we become more psychologically regenerative, inwardly rich enough to curb our other riches."*
> — Jack Adam Weber

That's the liberating truth—awareness hurts and that's okay.

You are much stronger than you think, and by choosing to stay actively aware of what's going on in the world, you are in control. I want to emphasize the *actively* because you will be aware regardless (that banging on the door will get hard to ignore). But with your eyes open instead of shut, you can take that information in controlled, bite-sized pieces. You can practice staying aware while keeping your feet steady on the ground. And with that awareness, you can begin to take empowered action.

If you *don't* do this, if you keep sleepwalking with a pink cloud around your head thinking that it's the only way to stay optimistic, you'll soon trip on something you can't see, and you'll fall.

I fell many times, and it was never fun. As you know, I had tantrums in the car, cried in the shower, and had days when I felt so down that I couldn't find reasons to leave bed at all. The pink cloud certainly wasn't working.

We know the world is messed up and many things are going in a very wrong direction. We know that species are becoming extinct at an alarming rate, that we're reaching atmospheric tipping points all over the world, and that greed and mindless consumer habits have created a system where we care more about money and fame than life and death. We know this—I know you know this—so how do we stay aware and still find hope to make a change?

We grow our hearts even bigger. That's the only way.

Have you ever had a moment when your heart grew so big, you honestly didn't know if it would hold or burst? Have you felt so much love, joy, or gratitude you didn't know what to do with yourself? You may have felt it at your graduation, while celebrating an end of an era with your friends and family there to support you. Or you may have felt it in small, intimate moments with the people you love. It's that feeling when the world inside you expands, and you start to wonder if you can take it, if your heart can hold this much happiness and love. You think it doesn't matter what happens next because at least you had this moment.

I want you to remember right now that your heart is capable of that love. Your heart can grow *so much*, in ways you didn't even think possible. For anyone who's become a parent (I'm about to learn about this experience and I can't wait), you may have had this happen to you. Love so strong you didn't even know you had it in you.

I know it's scary. I know that saying yes to love (no matter what kind) can lead to heartbreak and pain, and yes, the painful truth is that love *can* hurt. But just like leg muscles grow in pain, so does your heart. Keep giving it more, and it will keep growing bigger, stronger, and more capable of love. In other words, don't hold back on life. Be here for all of it or you might be missing out on the reasons to actually believe in a better world.

However, just because it's okay for awareness to hurt doesn't mean you have to keep your gates open to all the pain. You are only human after all, so know your limits. Later, we will talk about controlling your negative news intake, how to deal with what you learn, and how to grow emotional resilience so you can stay aware, yet empowered and motivated to take action.

You are made of 70 percent water.
The Earth's surface is made up of 70 percent water.
We are water.

So who are you not to let your teardrops flow?

Let. Yourself. Feel.

CHAPTER 38
Awareness Overload

Tonight, you decide, is the night you're finally going to watch that documentary everyone's talking about. You've been putting it off because you're nervous about how it'll make you feel, but everyone talking has made you curious—you want to be in the know too. You sit yourself down with a bowl of popcorn and press play. Let's learn about plastics in our oceans. Soon enough, you find yourself consumed—*how are things this bad?* As the popcorn shrinks, your awareness grows, and you feel yourself slowly changing. If you were living in oblivion before, you're not going to be that person anymore. This mindless dumping of our waste has got to come to an end.

Ninety minutes later, you go to bed with an aching heart and images of turtles with plastic straws in their bleeding noses. You can't believe we've come to this, and knowing this pains you, but at least you feel empowered, and empowered means you can take action, so you doze off with a new fire burning inside you—things have got to change.

The next morning, you wake up with images from last night's documentary still vivid in your memory. You feel changed, and you're determined to do what you can to fight this plastic pandemic and bring justice to the oceans. You can't wait to share what you've just learned so when you get to work or school, you talk about it with your colleagues or friends. Do they know how bad things are? And do they realize how much our daily habits are polluting the Earth and killing innocent life all over the planet?

Let's say they do know because they've seen the film too, and they actually have more insight than you because they've taken some time to look into the issue. And as the conversation goes, you learn it's not just about you and your daily habits, but about companies choosing to pollute despite environmental threats, and about politics allowing these acts in the first place. The more you talk about it, the more you understand how complex and complicated the issue is, and slowly, your commitment to saving the planet fades. Who are *you* to make a difference? As such a small fish in this huge ocean of power, who are you to make waves?

Before you know it, you hit the wall of awareness overload, and it doesn't matter what you try because you feel like a useless piece of plastic in the wind, unable to do anything but exist and get stuck on high branches.

Slowly, the images of the documentary fade as your normal life begins to take over, and you silently pray someone—an activist group, a bold politician, a company—will do something to end this mess. You'll hope someone else will do what you can't. Then you rationalize. You're just part of the system, and there's not much you can actually do without the system changing first. Maybe you go on hating yourself for every polluting action, or maybe you don't, but your inability to act is still there. The awareness overload has gotten the best of you.

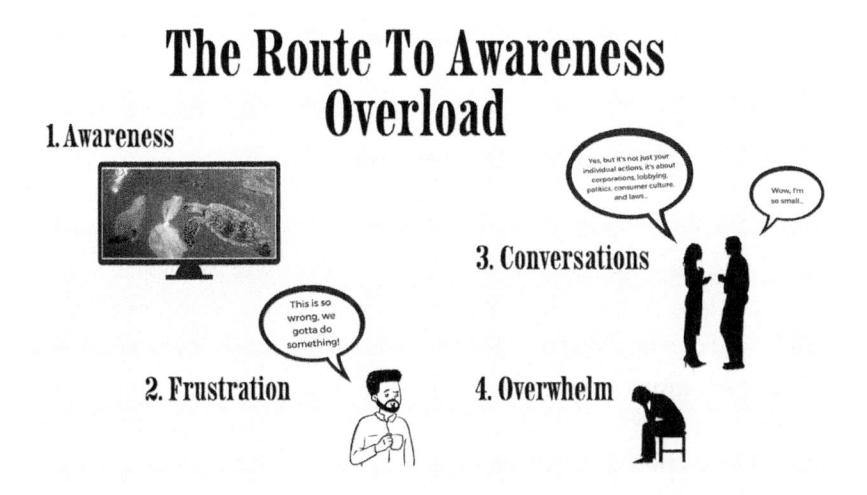

The Route To Awareness Overload

1. Awareness

2. Frustration

3. Conversations

4. Overwhelm

This is what I call the route to awareness overload. Maybe you've been here before. Maybe you've been here many times. Or maybe you haven't found yourself here quite yet. It doesn't matter. Chances are you will end up here in the future, and when you do, it's important to understand *why* you're here.

First of all, as discussed previously, there is way too much information out there for us to keep up with. With so much negative information at our fingertips, who doesn't get overwhelmed? Recognize it is a normal, very human reaction.

Second, companies in power who continue knowingly and willingly to pollute the earth want us to feel overwhelmed. If we are, they can continue doing what they're doing without interference. If you feel we can't do anything, they will happily play the inertia card and confirm your feelings. For example, it's been put on us—the consumer—to responsibly recycle packaging and products, although most of those products aren't even *actually* recyclable in the first place. Ludicrous? Yes. Yet this is how the world works. If a product exists that can't be recycled or harmlessly returned to the Earth, is it the company's fault for producing it, or the consumer's for buying it?

You get my point.

Unfortunately, that is the reality with many things in our world today, so it's important we don't get stuck in the overload phase for too long. Although *giants* are out there (companies with lots of money and power) who are on the other team, we can still win. As consumers and citizens, we have so much more power than we think, and the sooner we can get out of the paralyzing overload phase, the better.

Begin by giving yourself a break. Traveling the route to awareness overload is both common and normal, and we all do it from time to time—sustainability newbies and seasoned climate scientists alike. What's important is to ensure we don't get stuck there, that we start to take small steps out of overload and into empowered action because it is the only way we will ever actually make a change.

Yes, systems need to change, companies need to step up and take responsibility, and governments need to show they care more about life and the future of our planet than money. But that doesn't mean you don't have a part to play. As a citizen, you *are* the system, and every day, you make many, many choices that can build up to the changes you wish to see.

As a consumer, you can vote with your dollars and support companies with missions you believe in. As a citizen, you can get involved in local politics and show your elected officials what you believe in. You can call in and share your support for ongoing bills (something I highly recommend doing because it's *so* empowering) and you can also show up to public meetings and hearings. Gather some friends and family and your influence can suddenly become substantial.

And finally, as an activist, you can sign petitions (they matter); show up for a beach, park, stream, or road cleanup; help plant trees; join a community garden; or demonstrate against polluting activities in your area, to name a few. In other words, never underestimate your power to make a change.

Awareness isn't the issue. Awareness devoid of action is. This means continuing to shed light and spark awareness isn't going to do much unless we quickly move on to the next step—*doing something*. By slowly integrating yourself into powerful, positive action—no matter how small—you keep clear of overload, remain empowered and optimistic, and, ultimately, provide hope for others too.

CHAPTER 39
Climate Anxiety

"Perhaps the way to being the revolution is to stand up for your depression!"
— *Cultural Psychologist James Hillman*

It was 2006, and Al Gore's *An Inconvenient Truth* had just come out. I remember watching it a second time with my parents. I saw it in school first, then again at home, and as the credits rolled down the screen, I remember thinking, "The world is about to change." I was terrified by the findings, but grateful we knew what was going on. Luckily, we had a chance to act.

Keep in mind I was only fifteen and my understanding of the world was still forming. Therefore, I honestly thought the whole world would come together in a big room (like I'd seen things happen in the movies) and make decisions on how to save ourselves from this human-made disaster. I was expecting to read about it in the newspaper every day and to talk about it in school between sessions. With this much change ahead, I expected nothing short of a worldwide crisis, and surely nothing would go by unnoticed or unsaid.

Young and naive, I didn't understand the world was full of sharks who would do everything in their power to deny and spread false information about climate change so they could continue to make money. I also didn't understand that an undertaking like redesigning the world to fight climate change would take years, if not decades, or that we needed a whole lot of people in power to find the courage to do the right thing. In fact, I didn't understand much of any of this, and the only thing that soon became clear to me was none of what I thought would happen, would happen. No headlines, no global oval table crisis meetings, and no war-like crisis mode were on the horizon. It was just….more of the same. School went on as usual. My parents' lives went on as usual, and the world kept spinning with the same conversations we've always had. Nothing changed.

The only difference was *I knew*.

I tried doing what the rest of the planet seemed to be doing. I tried living a normal life and hoped someone in power was working on a plan. As a fifteen-year-old, I didn't think I had any power to do anything anyway, so I did what I thought was the right thing—I focused on school, grades, parties, and friends. I returned to being a typical teenager and tried to ignore climate change as much as possible; the thought of it was simply too hard to handle.

I didn't know then that by ignoring my fears, I started a fifteen-year bout of climate anxiety.

Climate anxiety is no joke. Having dealt with it for so long, I know this for certain. As I said, it caused rage, sobs, and tons of frustration, but it also grew deeper than all that. When I began healing myself in my late twenties, I recognized even the eating disorder and body dysmorphia I had struggled with my entire young adult life were rooted in climate anxiety. Since I couldn't control the world or all the horrible things going on, I needed something I *could* control, something to tie my sanity to that I was able to manage. I realized my calorie-mania, frantic exercising, and constant calculating of my food-and-exercise balance was my way of staying in control. If I could ensure I didn't gain any fat, I was in control, and I needed that. I *desperately* needed that control.

It took me years to heal because I frankly didn't know I was dealing with climate anxiety, nor did I have any idea what it was. When I was a teenager, no one was talking about this, and I didn't think I had anywhere to turn for help. I didn't know I should be looking for help.

That climate anxiety is something many are aware of today is both great and concerning. It means those suffering from it can get help, but it also means it's become so common that it's a topic worth talking about.

Pennsylvania State University psychology professor Janet Smith describes climate anxiety as, "Something people feel more and more when they get closer to an anti-goal, meaning a negative result, like the destruction of the planet."[15] In other words, we experience anxiety when we feel

---

15 Christensen, Jen. *"Climate anxiety is real, but there's something you can do about it."* CNN. May 7,

like we lack control. When it comes to climate change, I think we can agree that's how most of us feel.

If you are aware of climate change, chances are you're suffering from some degree of anxiety, maybe without even knowing it. The stages of climate anxiety can vary from "concerned" to "traumatized," where it can go as far as questioning your reasons for staying alive. If you are at this stage, or if you find yourself losing sleep or focus due to climate stress, seek help. I'm sharing helpful ways of working through your anxiety and growing through your emotions, but I am not a health practitioner, so if you need medical attention, please seek it.

No matter where you are on the spectrum, understand that feeling anxious about climate change is normal. In fact, as sad as it is to say, 73 percent of Americans agree that climate change is probably or definitely affecting their mental health.[16]

Not only is the increasing epidemic of climate anxiety bad for our overall wellbeing, but it's taking a toll on our important work. As Smith explains, "Anxious people tend to shut down and not engage."

Although what we really need is to feel empowered and inspired to take action, the *opposite* is happening. Not only is the fear of climate change digging us deeper into the hole of despair, but it's also keeping us stuck there. Therefore, it is important we check in with ourselves (preferably regularly) to see how we are feeling.

If you are worried or feel fear or anger, allow yourself to tap into these feelings. Remember ignorance is not always bliss and that by remaining aware, you can work through your emotions and let them empower you instead of allowing them to be the gateway to despair.

At the end of this chapter, I will share some tips I've learned (for myself and from real experts in the field) that can help you with whatever climate anxiety you might have.

2019. https://edition.cnn.com/2019/05/07/health/climate-anxiety-eprise/index.html

16 "New APA Poll Reveals That Americans Are Increasingly Anxious About Climate Change's Impact on Planet, Mental Health." American Psychiatric Association. October 21, 2020. https://www.psychiatry.org/newsroom/news-releases/climate-poll-2020

Difference Between Climate Anxiety and Eco-Grief

Climate anxiety is what you suffer from when you are worried about climate change and the many scary scenarios this crisis could lead to. If you're a parent, you might be worried about your kids' future. If you're a teenager, you might very well be worried about your own. The reasons for climate anxiety can be many, and they are often attached to personal threats to your own wellbeing. Since the future is so uncertain due to this threat, the reasons to worry are endless, but more often than not, they are tied to our own fears.

Eco-grief is not quite the same, although the two are closely linked, and many who suffer from one, suffer from both. Eco-grief simply means you mourn the loss of our natural world.

In *What We Think About When We Try Not to Think About Global Warming*, Per Espen Stoknes explains we have three main sets of values: egoistic, altruistic, and biospheric. Egoistic values are those focused on personal achievements, like social power, wealth, and personal success. Altruistic values make us care about other people and the bigger community, the kinds of values that make us good citizens. Finally, biospheric values focus on the health of beings like plants, animals, trees, mountains, and other ecosystems—it's about caring for the living world.

Stoknes further explains we all have each of these values; however, they are weighted differently from person to person. Research confirms the more connected one is to nature, the stronger their biospheric values are. This means if you spent a great deal of time in nature as a child (or later as an adult), you feel a *connection* to nature. To you, nature is more than just a place to go from time to time. You *care* for nature as if it were a sibling or friend. You understand nature is part of you, as it is part of all of us, and coming to that realization can be the most healing feeling in the world.

But it can also be extremely devastating when you connect the dots and understand that, gulp, we're all dying. Ecocide is real, and when Mother Earth is hurting, so are we. Not only do we rely on ecosystems for our world to function, which makes the rapid decline of species around the world even more heartbreaking, but we feel this loss on a spiritual level too.[17]

---

[17] The rapid loss of species we're seeing today is estimated by experts to be between 1,000 and 10,000 higher than the natural extinction rate (The World Wide Fund for Nature: https://wwf.panda.org/discover/our_focus/biodiversity/biodiversity/). On average, we lose about 150 species per day (Yale Environment 360, blog published by the Yale School of the Environment, Fred Pearce, August 17, 2015: https://e360.yale.edu/features/global_extinction_rates_why_do_estimates_vary_so_wildly).

If you can relate to this, if you feel like the loss of nature pains you gravely, embrace that about yourself—it means you're deeply connected. You just have to learn how to use it as your strength and ensure we save whatever we still have left.

Since you likely have all these values—egoistic, altruistic, biospheric—there's a great chance you suffer climate depression to some degree, even if you're not aware of it. I recommend doing the journaling exercise below so you can gain a better understanding of your current emotional resilience.

One more thing—don't fear grief. We tend to have a strained relationship with grief like grieving is something we do behind closed doors when no one is watching so we can appear strong when we step back into the world again. But what if grief doesn't make us weaker, but *stronger*? What if it's the missing piece that will enable us to actually do something about this crisis we're in?

> *"There's a hidden secret in the despair paradox. Going down the depths of despair can also bring healing.... The more we let death—even the threats of extinction—into our souls, the more we can appreciate the current vibrant vitality of life in its many forms. And we may even be transformed by it."*
> *— Per Epsen Stoknes*

CHAPTER 40
Activate Healing

My climate optimist awakening saved me in many ways. It was when I recognized I had to start taking serious care of myself or none of my work would matter. If I'm not strong, how can I help the world? If I can't be light, how can I activate the light in others?

I'm grateful for the ups and downs my mental climate journey has taken me on because I've learned the only way to change the world is by changing ourselves first. And the only way to *heal* the world is by healing ourselves. If we don't choose to actively work on our own healing, how can we serve the healing of others? If we can't find our light, how will we help others find theirs?

If a friend falls down a well, you don't want to jump down after them and say, "Hi, I thought you might want some company down here. At least now we're stuck together." Do that, and they'll look at you in disbelief and simply reply, "Idiot." Instead, you want to find a rope, muster all your strength, and help them out. The same goes for our healing work.

Although empathy and the ability to see and understand someone else's pain is a beautiful act in the moment, you don't help them in the long run by lowering yourself to the same level of pain. If you want to truly help someone, let them grieve as needed, waiting to offer a portal back into the light. They won't want to be stuck in Sorrowland forever, and when they're ready, they will seek out the light again. You get to be that light if you choose to be a conduit for healing, an anchor to the good energy so needed in this world.

When it comes to the world at large, I believe it's important we understand this idea of keeping the light burning. With all the pain surrounding you, more often than not, you might feel that to be a good citizen, you should shoulder the fears and sorrows of everyone around you. If you don't show them you're heartbroken and disturbed all the time—because how could you not be with everything going on in the world?—it might come across like you don't care. Therefore, it's almost easier to be angry and worried all the time because it's what you think is expected. Choosing to channel the light can seem scary, but that is also how you begin embodying the change we so desperately need.

When I chose to hurt myself to "feel the pain" of all the animals around the world, I wasn't helping anyone. I recognize that now. But it took me many years of active healing to get where I am today. I still have days when it's hard to keep the tears down, and when those days come, I allow myself to cry because when I allow myself to feel, I also allow myself to heal.

I want you to become a master at this. I want you to get so good at staying attuned to your feelings and emotions that you can process them effortlessly when needed, and continue to be a conduit of healing and light for yourself and for the world.

This work is not a "one and done" thing. Healing and emotional empowerment take practice, and just like with working out, to stay strong, you need to commit to it for the long haul. But similar to working out, once you've reached a certain physique and made your new normal strong and resilient, sticking to these habits will become second nature for you. It will be who you are—a healing vessel for change.

To me, climate healing relies on four pillars:

1. Start by recognizing that it's *okay to feel* and let yourself be one with your emotions. Ignoring your feelings will not make them go away. They will only build up inside, which can lead to anxiety and depression. Letting yourself feel is the first step in healing. You might feel scared by this, but your heart is stronger than you know, and if you practice tuning in to your emotions more often, you will soon learn how to let them flow through you with ease.
2. Once you've recognized it's okay to feel your emotions, no matter how difficult they are, it's important to remember you don't have to hold on to these feelings to make a difference. You're not a better activist because you're angry and worried all the time. In fact, you'll

be much more effective if you choose to heal yourself and act from a place of strength, love, and light.
3. Understand that you're not alone. By talking about it with others, you activate your own healing, first of all, but you also allow others to start their healing process. When we share how we feel, we begin to understand we're not alone. That is a good first step in releasing pain. Also, when we come together in fear and grief, we feel like we have support, like there is hope. Hope is where the action grows.
4. This leads me to the last part—taking action. Professor Smith explains the fastest way out of anxiety is by feeling you have control. When you take action, you feel you have control. Whatever you can do to make a difference, no matter how small it may seem, do it. Become the change you wish to see, and slowly work yourself out of despair and into empowered action. (We'll talk about this a lot more in the Choosing Empowerment section.)

You can't grow abs by closing your eyes and dreaming about them. At some point, you have to get on the floor and do some crunches. The same goes for emotional resilience. You can't just tell yourself you want to feel better. You have to seek out activities that activate healing.

I prefer free, simple, and easy to access at all times activities. My favorite exercises for mental health are journaling, talking, and positive action. Here are some resources you can use to activate your own healing today.

Healing Journaling Exercises

Journaling could be one of the fastest ways to discover hidden emotions and release the energy that is blocking you. I recommend it for any sort of deeper reflection, or as a daily practice to tune into yourself and understand what's going on below the surface. We are our emotions, after all, and they tend to be behind the vast majority of our decisions, so the more in tune with our emotions we are, the more empowered we will be.

Journaling is supposed to be your own sacred ritual, and whatever feels best for you is the key here. Maybe you prefer to journal first thing in the morning, at lunch, or right before bed. Figure out what works for you, making sure it's something you enjoy doing.

Although journaling is good for all things, these four guides will be specifically focused on eco-grief and climate anxiety. Use them as a tool to acknowledge whatever grief or anxiety you might be holding onto, work them through your body, and finally release them so you can heal. Develop a regular journaling practice or turn to it whenever you feel you need it. The more you work on emotional resilience, the stronger and more motivated you'll feel as a climate activist and earth hero.

Part One—Landing in Grief

This is an excellent exercise for growing emotional resilience.

1. Write a list of all the ways humans are killing/hurting/destroying our planet and the natural world. Write as much or as little as you can think of.
2. On a new page, write down how this makes you feel. Be colorful—swear if you have to. Consider it a hate letter to the system. Don't worry—no one but you has to read it.
3. Read (out loud if you can) what you just wrote. How does it make you feel? Breathe deeply and allow yourself to really feel whatever comes to mind. Don't hold back. If you feel like crying, cry. If you feel like yelling or punching something, please try to find a (safe) way to do so. By letting your emotions process through your physical body, you are releasing them.

Part Two—Releasing and Healing: The Self

In this part, I want you to recognize yourself as part of the pain. Who are you in all of this? And how do you feel you're not being seen, heard, or taken seriously in this work? Part of our ecological grief and climate anxiety is tied to the feeling that we aren't enough, or we don't have enough power to make a difference. That is hurtful and disempowering in itself, and it certainly isn't true. This session is about healing yourself as part of the ecosystem.

1. Write about times when you weren't heard or when your message wasn't received. How did that make you feel? Or, if you can't think of any, list the ways you feel small and disempowered within the

big, complex issue that is climate change.
2. Now, rewrite the story. What if you were strong/wise/smart/powerful enough to make a difference? How would your message be received? How would you make a difference? And how would receiving this respect—from yourself and others—make you feel? Write it down.
3. Read what you wrote to yourself. How do you feel? How would you feel if you got to wake up in the morning and feel powerful enough to make a difference?
4. Close your eyes and see yourself in that light—you are that person. Feel it, believe it, and see it. Understand you are changing the world, and you will continue to do so every single day.

Part Three—Returning to Light

Our hearts are broken over the world and for good reasons, but how can we ever wake up with enthusiasm if we don't mend our relationship first? We're stuck in a relationship that no longer works, and we're too heartbroken to admit our mistakes. It's about time we come to terms with that so we can heal and then create something even better.

Part Three is all about moving forward. What future could we make if we were to heal our wounds and move on? What is possible if we wiped the slate clean and started over with happy and enthusiastic hearts?

1. Take a deep breath, and feel every muscle in your body easing up. Close your eyes and invite the light in. Ask it to heal and wash away your sorrows. Forgive yourself and humanity for all we've done to disrespect the earth and each other. Tell the light you're ready to move on.
2. Now, transport yourself twenty years into the future where changes beyond your wildest imagination have taken shape. What do you see? What would it look like if we healed all our broken relationships and moved on?
3. Write this future down. Describe it in detail. What did you see? Nothing has to make sense. Just write down whatever comes to you, as unrealistic as it might seem.

4. Read the story to yourself—how does it make you feel? If you feel expanded and optimistic, does it feel *okay* to feel that way? Practice being okay with this feeling.

Part Four—Reimagine Yourself

In Parts One and Two, we released the fear and grief we were holding on to. First to the world and then to ourselves. In Part Three we moved on to envisioning the world we want to see if we were to dream without limits. I hope you created a beautiful world for all of us to live in.

Now I want to turn the spotlight back onto you. If you were to go deep into yourself where everything is possible with no limiting beliefs, who would you be? What would you be doing in this world, and how would you make a difference?

Sometimes it's hard to trust our guided paths because the world we currently live in doesn't support that journey. You might have an inkling of what you want to do, but your dream stops abruptly when you realize you can't make money doing what you desire (or at least you think so). I want this exercise to be about creating space for yourself without limits. I want you to write as if everything were possible, as if no dream or desire were too unreasonable—because who says it is?

1. Take a deep breath and close your eyes. Go to the future you wrote about in Part Three and bathe in that light and energy for a moment.
2. Use this energy to channel your inner knowing and ask, "What am I here to do?" Write down all that comes to mind. Where do you live? What do you do? How are you making a difference in the world? Write as detailed an account as you can, and have fun describing the life of your dreams.
3. Read it out loud to yourself. How does it feel? Can you bring yourself to believe it?
4. Remember, when you start to manifest new things, it can feel uncomfortable at first. Or you might feel like you need to know all the steps to getting there. You don't. All you have to do right now is have a strong vision of the future you want to see and then practice having the faith and energy to believe in it. Allow yourself to dream.

These guided journaling exercises were created to help you shift the narrative of your stories from hopelessness and despair to healing, empowerment, and excitement. Do them if they speak to you; ignore them if they don't. As long as you find ways to heal yourself regularly, you can do whatever works best for you.

Talk About It

Katharine Hayhoe, a renowned climate scientist and one of the most optimistic people I've ever had the honor to talk to, says the most important thing we can do about the climate crisis is talk about it.[18] By normalizing these conversations, it will become evident it's something we all care about—regardless of religion, faith, culture, or background—and we can come together as a species and get this right. But it's not just about making people climate aware. Talking about our climate fears is also an excellent tool for healing.

When we talk about something with others, we open up a powerful force—connection. We see we're not alone in our feelings. That, in itself, is a big step toward healing. How often do we carry a heavy load of anxiety and fear thinking we're the *only one in the whole wide world* feeling this way? I'm not sure if that's you, but I've definitely been there way too many times.

My mom often told me, "Honey, you can't carry the world on your shoulders." Of course, she was right. I didn't know then that I didn't have to—or that there were so many others passionately seeking change. If we connect with one another and recognize this shared pain, we give ourselves a little breather. It's like, "Phew, I'm not alone," and then the action can follow.

However, don't start a conversation thinking you have to find the solution by the end of it. When it comes to climate change, I'm not sure all the solutions even exist yet.

I don't know when we decided if we didn't have a solution, we shouldn't voice our thoughts—don't you dare *complain*. We live with the dangerous stigma that it's not acceptable to complain about current

---

[18] Read Hayhoe's book Saving Us: *A Climate Scientist's Case for Hope and Healing in a Divided World* if you want to learn more about how to start empowering and change-making conversations.

circumstances unless you have something valuable to bring to the table. Therefore, we're told to suck it up and go about our days as if nothing is wrong—while the fear inside grows too hard to deal with.

Only one in four Americans hear someone they know talk about climate change every month. Considering it's the biggest threat to the global economy, that 9 million people die from air pollution alone each year, and that scientists have given us less than ten years to do something serious or it'll be too late, you'd think it would come up more often.

Why doesn't it?

Because it's steeped in shame. Because we're afraid to *feel*. And because I think we're all holding on to some degree of climate anxiety, and we don't know how to handle it. We have too much other shit going on as it is.

It's okay to bitch about climate change. It's okay to tell the people around you that you're worried about the state of the world and the outlook for our future. It's okay to feel anger and despair with the world without being part of an organization working to make it better. Truly—it's absolutely fine to feel bad about our warming planet, and at the same time, recognize you're part of the problem. We have to come back to ground zero and give ourselves a break. "Yes, I know I'm a huge contributor to climate change, but that doesn't mean I'm comfortable with where this is going. And I am allowed to be worried, although I don't know how to make this right."

By speaking up and sharing your thoughts with the people around you, you invite 360-degree healing. You begin to heal yourself while allowing others to tap into their awareness and grieving as well. Don't keep it to yourself. Be that bold person who dares to wear their heart on their sleeve and tell the world you're hurting. We must make it normal to talk about climate change, and that starts with trust. It starts with unity and mutual understanding, and by telling others it's okay to be worried, angry, sad—even ashamed. Joanna Macy, environmental activist, author, and scholar of Buddhism, general systems theory, and deep ecology, says it so beautifully: "It's okay for the heart to be broken over the world. What else is a heart for?"

CHAPTER 41
What Else Is a Heart For?

I ended the last chapter with one of the most beautiful quotes I've ever come across. It was spoken by the even more beautiful human, Joanna Macy. Listening to her speak on a podcast a few years back helped me gain a whole new understanding of the capacity of my own heart.[19] Something she said, a parable she shared with the listeners, struck a chord, and I haven't stopped thinking about it since. Every time I start to question my ability to go on with this work, I come back to what she said, and it always helps me find new light and move forward.

I wrote the story from memory so it is slightly different, but the message remains.

What she said was:

Indent next paragraph as quote

Say you're taking care of your mother, and she's dying of cancer. You won't say, "I can't go in her house or in her room because I don't want to look at her." If you love her, you want to be with her. When you love something, your love doesn't say, "Well, too bad my kid has leukemia, so I won't go near her." It's just the opposite.

Inspired by her wise words, I will paint the scenario further.

A young parent sits by a hospital bed. Their child is dying from cancer. Never did they think their life would come to this, but there they are, looking this painful reality right in the face. It is by far the most difficult thing they have ever gone through. They wish no one would ever have to face this reality again. They wish things were different. That they were out running in a field instead of stuck in this stuffy hospital room. They wish they were blowing gently on their child's scratch from falling, not looking at tubes

---

[19] "Joanna Macy: A Wild Love for the World." Episode on *On Being Podcast* with Krista Tippett, April 25, 2019.

coming out of their child that are keeping their precious angel alive.

Will the young parent love their child less because the cancer is too ugly to see? Or will they, if even possible as a parent, love their child even more? Will they close their eyes to this pain and walk away, or will they sit there, day in and day out, praying every possible prayer that their child will be okay?

Of course, you know the answer—they will grow their heart so big, their faith so strong, and their will to believe so unbreakable, it will seem like there is nothing in the world they cannot do. They will try. They will talk to every doctor, reach out to every expert, and seek support from every colleague and friend—anything they can do to save their precious child. That is what a parent would do. They would love so much there's no thought of ever backing down. And they will be there through all of it. Whatever there is to let in—the good and the bad, the ugly and the pretty—they will let it in. They will be there by the bedside until their child's last breath, and they will not give up. As long as there is hope, the young parent will do whatever they can to fight, and their love will never falter. In fact, love is all that keeps them alive.

Joanna Macy's point was that just as you would love and grieve a sick child openly with no shame or reservations, you must know you're allowed to love and grieve a broken world. How can we find the courage—or the motivation—to keep fighting if we don't? How will we ever know what we're losing if we don't allow ourselves to feel it?

Yes, there is a lot of heartbreak and suffering all over this world, and more often than not, we look the other way. We'd rather be insulated from the pain because we believe our hearts can't take it. But just as a grieving parent's heart can grow beyond belief, so can yours. You can give yourself permission to activate that growth right now, not so you can feel more pain, but so you can find the strength to fight for the light.

The only thing that can truly save us, more so than technology, politics, renewable energy, and laws, is our growing hearts. Our hearts recognize there is something worth caring for in the first place. Our hearts can not only handle hard things, but grow stronger and more resilient from our pain. Our hearts, if activated, can become beautiful vessels for healing and continue to fight for hope as long as they still beat.

This is what we have to put our faith in—our beautiful, strong, and beating hearts. If you take anything away from reading this book, I hope it's this: It's okay to feel it all. What else is a heart for?

CHAPTER 42
Raising Your Fear Mark

On a scale from one to ten, what is your fear mark? I bet no one has asked you this question before. The more I think about it, the more absurd I find it that we don't ask more often. How is it not one of the top ten questions we ask on a first date? Why are we not practicing our response to that question when preparing for a job interview? Wouldn't you want to know how skilled and resilient a potential future employee or partner is when it comes to dealing with shit when shit gets hard?

I didn't know I had a fear mark until I read *Climate Cure: Heal Yourself and Heal the Planet* by Jack Adam Weber and found myself having aha moments. I realized then that I've never reflected on what my ability is to take in difficult information nor that I should practice how to raise that ability. Although I've learned awareness hurts and that's okay, I haven't actually reflected on *how I get better* at feeling that way. Interesting.

Learning how to deal with negative news is something they don't teach you in school but absolutely should. Instead, it seems we get an overload of incredibly difficult information dumped on us without any help or guidance on how to actually *deal with it*. It's like we've forgotten that behind all our titles, outfits, and social media accounts, we're actually humans. We are living, breathing things with pumping blood and beating hearts, bodies with *feelings* and minds that process information way beyond what any robot would have. (I know robots are becoming astonishingly smart these days with cute "feelings" and all, but stay with me here.)

If you're anything like me, you may have experienced compassion overload. It gets to the point when you simply can't see another story about wildfires, oil spills, or school shootings—it's too much to take in. You reach a point when you *simply can't care anymore*. The world seems to be going under, and you feel there's nothing you can do but surrender to that fate.

If we're going under, so be it—I'm tapping out.

This is what happens when we run out of emotional storage. In *Climate Cure*, Weber call it our fear mark: "Our fear mark is the degree of fear we can tolerate while remaining rational and skillful in our response to information. Until we develop awareness, our fears rule us."

Until we develop awareness, our fears rule us. In other words, it's not enough to be aware of the issue at hand. We also need to be aware of how it might make us *feel*. If we're blind to the fact that learning certain things hurts, we are powerless to properly handle the information. Raising our fear mark, then, means getting better at staying aware without getting overwhelmed.

Chances are if you're not aware of your fear mark or haven't been trained to raise it, you can't take that much. Someone you meet on the street could say something you're not ready to hear and your immediate response is, "Oh, I know, it's terrible. I don't want to talk about it."

That's fair. We don't have to talk about everything that's going on in the world (and I'm getting to that point shortly), but if our instant reaction is we don't want to know, we're keeping ourselves stuck in our current state of awareness. From there, it's impossible to grow.

If you don't understand your fear mark, you probably have a low threshold for bad news. You might *hear* the news, but you're not letting it actually touch you. If your fear mark is low, it's simply too hard to handle difficult information. Instead, you close those ports and hope the painful feeling will pass by you and move on.

But we need to be aware, so ensuring we can handle difficult information is paramount. It's not enough to say we're ready. We have to actually *be* ready. That is where acknowledging our fear mark comes in. Because once we acknowledge it, we can begin to grow it.

In the beginning, your ability to take in negative news might be low, and that's okay. There's no need to overwhelm yourself, and by taking in bad news in bite-size pieces, you can get stronger and raise your fear mark over time. And if you can raise it really high, you can know what's going on in the world, actually *feel* it in your emotional body, without letting it overwhelm you. You'll be a super-compassionate being with the power to—in all honesty—change the world.

One of those super-compassionate people was Mother Teresa. I read she had to ignore homeless people on the street daily. It surprised me. I thought she would be so aware she could never ignore someone in need. But then I understood she *did* see, but she also recognized she couldn't help everyone. And part of being vividly aware and strong was being able to understand when you should act and when you should not.

If we try to be everything for everyone, we'll end up being nothing to anyone.

I live in New York. When I first moved here, I found it so difficult to go anywhere because whether you take the subway or walk down the street, you will pass by someone in need. Being a poor student at the time, it was difficult to know when to drop a dollar in their cup. I felt so ashamed and disappointed with myself. Who was I to ignore suffering?

But then I remembered what I had read about Mother Teresa and reminded myself not every battle was mine to fight. Simply recognizing others don't have it as good as me is sometimes enough. I don't have enough money to feed every person experiencing homelessness I pass. That's just a fact. I can choose to acknowledge I can't help everyone, or I can close my eyes and ignore the people suffering all around me. One requires strength and humility; the other is the easy way out. I realized that choosing not to see the suffering would be worse than acknowledging my economic shortcomings. It takes strength to see something, *want* to help, and at the same time accept that you can't always do so.

A couple of years later, I attended a practice run at a secret theater club where a group of acting students performed a sketch as if they were homeless. To practice, they had spent a couple of nights on the street with actual people experiencing homelessness (I was very impressed), and they said what they learned stunned them.

One man had told them something in particular that changed how they view people experiencing homelessness forever, and by sharing it on the stage, they did the same for me. This man had said what he was asking for, more than anything else, was to be seen. "It's okay if you can't give money or food," he had said. "We can understand that. But if you can just acknowledge that we're there, maybe give a nod and a smile if you feel comfortable doing so, that's enough." People experiencing homelessness feel their absolute worst when people pass by and act as if they aren't even there.

What that teaches us is that the most important thing we have in this world is the love and energy shared between people—known or unknown—and never to take the relationships we share for granted. (Think about that the next time someone yells at you for something. At least they recognize your existence!) I think it is also a great reminder that it's okay to acknowledge difficulty and pain and tell ourselves we can't always step in and help. Think of Mother Teresa, the saint of all saints, who walked by the homeless and recognized she couldn't help them all individually. If she could do that, so can you, but you have to raise your fear mark first. You have to learn to get comfortable with the world as it is before you can fully take your place in it.

Here are five steps for how to deal with difficult and negative news I learned when I co-taught one of the Climate Optimist Master Classes with my friend, Jack Adam Weber. Practice these and you will notice your emotional resilience grow and your fear mark with it.

1. **Find your balance.** Too little awareness might lead to anxiety and stress, and too much can overwhelm you. Recognize that we're all on different resilience levels and some people can consume more negative news than others. By knowing yourself and your limits, you can begin to recognize when you feel overwhelmed and take a break. Keep it bitesize and grow your emotional resilience over time.
2. **Practice healthy denial.** We all deny negative news to one degree or another. It would simply be impossible to live and function if we were to think about it all the time. Denial isn't always bad, and tuning out from time to time is essential for our wellbeing.
3. **Mix in optimism.** A good way to consume negative news is to follow up with something positive. That way you can remain aware without feeling like you lose yourself to anxiety and stress. It's okay to find reasons to smile amid all the bad and *you*—the everyday

hero choosing to work on your emotional resilience—deserve it more than anyone. Seek out the good news too. Listen to uplifting music. Sing, dance, and enjoy life. Do that and you'll be much better equipped to do something about the things you're learning.

4. **Talk about it.** By sharing what you're learning with others, you might find that you're not alone in your worries, which will help ease anxiety. Just keep in mind that others' emotional resilience might be very different from yours—tread lightly and with kindness.

5. **Fear + Grief = Empowered Action.** Grief alone can consume you. Fear alone can push you into a paralyzed state. But when combined in healthy doses, and applied to an attitude of wanting to make change, you can fuel it into empowered action. Don't just learn and let go—empower yourself and do something about it. Knowing what you know now, what can you do today to make it better? Choose to act and be the change, and be assured the world will follow.

CHAPTER 43
Dealing with Negative Climate News

You just learned about the importance of raising your fear mark. A huge part of that work comes from learning how to consume negative news. If you lack this skill, it's easy to lose yourself to either despair or denial. Either you become so consumed by the negative you lose all hope, or you simply shut the awareness door and dive into things (distractions) that are easier to handle. Both outcomes could turn out to be deadly. (Excuse the drama, but it's true.)

For starters, remaining aware of the state of the world could be more critical than ever before, not just because climate change is a threat beyond any previously seen, but because it will, if it has not already, affect all of us.

I recently learned that if we pass 2.0 degree Celsius (or 3.6 degrees Fahrenheit) of warming this century (the target is to limit it to a 1.5-degree per the 2015 Paris climate agreement), many, if not most, insurance companies will say "no more." It simply wouldn't be financially reasonable—*or possible*—to offer insurance in a world that is so certain.[20] In other words, we'd be looking at an uninsurable world, and we're headed there faster than we can imagine. If you're a homeowner or looking to become one, be aware.

Second, if we lose ourselves to either denial or despair, we miss the opportunity to participate in the changes. Lost without direction or hope, it is incredibly hard to take action or even think that it matters if we do. This is why the brilliant thought leader and author Yuval Noah Harari said: *"Climate despair is as dangerous as climate denial."* They both keep us stuck right where we are. Therefore, learning how to be smart and kind to ourselves about our news intake is critical, both for our mental health and for the world.

---

20 Irfan, Umair. "The $5 trillion insurance industry faces a reckoning. Blame climate change." October 15, 2021. https://www.vox.com/22686124/climate-change-insurance-flood-wildfire-hurricane-risk

Many of the following tips are similar to the ones under raising your fear mark. But since the two go hand-in-hand, they very much apply to both.

One—Get Mindful and Filter Out

You don't have to read every article or watch every video on climate change. If you find yourself overwhelmed by climate news and feel like you can't get a break, start by simply cutting down. If it gets to be too much, it's hard to process any at all, so try finding a healthy balance that works for you. Consider signing up for a weekly newsletter or a news platform that gives you daily tidbits. That way you can stay informed, but not get overwhelmed.

Also, if you come across a video on social media or an article that catches your attention, but you don't feel like you're in a good place (physically or mentally) to take it in, make a note and come back to it later. Nature is circular, and so are you. There will be times when you're more vulnerable and not up to handling "attacks" from the outside world. Honor this. Be kind to yourself and know your limits.

Two—Follow Up with Positive News

Try to follow up negative news with some positive stories. We don't hear about them often, but there are some remarkable climate stories out there worth keeping an eye on. Reading good climate news to balance the negative is essential for keeping optimism alive. It makes you more resilient in your work for climate justice, and it also helps you recognize and celebrate all the incredible solutions already underway.

It might seem like we have a mountain to climb to figure this out, and we do, but by recognizing we've already begun the climb, it's much easier to find the hope, the optimism to keep going.

Three—Healthy Denial

Know that it's perfectly fine to apply a bit of healthy denial. We all live with denial to one degree or another. It would be quite impossible to function in our everyday lives if we didn't, especially given how easy it is to access information.

As you know, awareness overload is a real thing, and it can easily lead to suppressed feelings, anxiety, and most likely, inaction. Therefore, it's actually part of your mission as a climate optimist to be mindful of how much you can consume while staying optimistic and motivated to act.

How you apply this denial is up to you, but make sure to take breaks during your day when you *don't* think about climate change. Go for a walk, play with your kids, go have a date with your friend. There is so much to rejoice about, so much beauty to appreciate, so make sure you allow yourself to do so. Find time to simply *live* to keep your strength (and sanity) moving forward.

Four—Community

Any burden is easier carried if shared, so the best way to deal with climate anxiety and overload is by finding support from others. Start having conversations and share how you feel, even if you don't quite know how to put words to those feelings. Who knows? Maybe your friend, parent, or colleague feels the same.

It's okay to be worried about climate change, and it's important we openly share that we are. That way, we allow others to feel the same way, and we can come together to enact powerful changes.

I also highly recommend joining a community of like-minded people. Then you get to share your fears with people who understand you but also take positive action. Find a community physically close to you or join one online. Many resources are available to support groups and positive climate change action groups, so don't for a second feel like you're in this alone. Look for any community where you get to share your fears but also participate in the positive action.

These were just a few tips for getting better at consuming climate news, but of course, find what works best for you. Just understand that more knowledge does not necessarily lead to more action. Being balanced and grounded is key if you want to be an empowered ally, so keep that in mind. You don't have to know *everything* that's going on. You can also start easy and slowly build up to a higher tolerance. As you practice the steps above, you will learn it's much easier to be informed while not losing hope, since the mindful practice and the action itself will keep you balanced.

And if you don't find a healthy balance right away, go easy on yourself. Maybe start by taking a media break altogether and build up some new strength. Once you feel more balanced, start adding a few news snippets into your daily routine again. Follow the tips above, with limited consumption followed by good news and remind yourself that a bit of healthy denial is totally okay.

CHAPTER 44
Should I Have Kids?

Writing this at four months pregnant, it might be hard for me to be objective about whether or not one should have kids. However, this question has lived with me for a long time, and I know I haven't been alone in worrying about bringing children into this world. For one, how can we bring new life into a world that is so messy? How can we, with a good conscience, birth another living being into this place? The kids alive today are already terrified about their future, so how bad will things be when my kid grows up?

I totally understand the sentiment of not wanting to bring that kind of pain and uncertainty on another human, especially not your own child, so if you're carrying these thoughts, I feel you. Sadly enough, four out of ten young people fear having kids due to climate change, which tells us something about where we're headed.[21] If our next generation doesn't want to continue human life on Earth, what does that say about us? How messed up has our world actually become?

This might please many climate nihilists who think the human race is the source of all evil. These people will argue that what we're really facing is a population issue, and we simply need to stop having kids.

We're not facing a population issue. We're facing a human values and capitalism issue, an issue where a very few pollute more in a day than the rest of us do our entire life. That being said, I am definitely part of the percentage of the population that pollutes too much when people in underdeveloped countries have to pay the highest price. Those people have done the least to fuel climate change but suffer the most. It is far from fair, and *that* is what we need to pay attention to. We need to find ways to live

---

21 Harvey, Fiona. "Four in 10 young people fear having children due to climate crisis." *The Guardian*. September 14, 2021.

with less of a negative footprint while learning to maximize our positive one. That is the issue we're facing, not whether or not we should have kids.

We have enough resources on Earth to feed every human and electrify every home. Resources are not the issue—how we distribute them is. Therefore, it's not a population issue *per se*, but a very skewed idea of how life on Planet Earth ought to be lived.

As described earlier, becoming a good steward of the Earth is not about sacrificing all the things we've worked so hard for. Instead, it's about continuing to grow, expand, and find even better ways. We must find the courage within us to question the values we live by now and explore how things could be even better. Maybe less in many ways, but better nonetheless.

Not having kids would mean going against our very nature as a species and saying no to reproduction. As a woman, I would have to turn off my internal clock and ignore my body's longing to become a mother. I would have to say, "Shame on you for wanting this," and shut down that part of myself. I would have to become a little less of a woman, and in doing so, become a little less human.

I don't think we need to become less human to save ourselves. If anything, we have to become *more* human. We have to reconnect with our deeper knowing, our inner compass that knows how to live beautifully here on Earth. We must respect our body's deepest desires and understand that when we do, things will fall into place.

If we could listen to our bodies when they're tired and simply slow down, we could regain balance and fall into our natural rhythm of life. If we can live more grounded and attuned to ourselves and our feelings, we don't need as much to keep us happy, healthy, and strong. No medicine, energy drink, or shopping trip can substitute for what a life of harmony and balance can do for us. And when we do find harmony, we're not the problem on Earth—we're part of the solution that will continue to make this world a more beautiful place.

Sure, every human comes with a footprint, more so in the Western world than anywhere else. As an expectant mother, I know very well there's an entire market out there telling you what your baby needs. And that's only for its first year of life. Who knows the trash pile this human will leave behind as they get older? As with anything, our existence can become a huge negative bomb, but it can also serve as a beautiful opportunity to make a difference in the world. Every soul belongs to the universal song and, therefore, can become a tune so beautiful everyone will want to listen. In other, not-so-metaphorical words, your child can be one of those leaders who helps the world move forward.

If you want to become a parent, recognize that you have the beautiful opportunity to bring a powerful ally into the world. You get to influence this little nugget to learn the values you want to live by, and together you get to travel this exciting journey toward a better tomorrow. If we say no to having kids, we have given up on our future altogether, and that is not what a climate optimist would do. Upon realizing this, I finally gave in to my heart's calling and said yes to growing a new life, and it's with the deepest smile my body can muster that I say this—I am so excited to have a child!

CHAPTER 45
Let Your Spirit Catch Up

I'm very good at being busy just to be busy. I started working at a local coffee shop when I was fourteen and kept it as a side hustle while I climbed through school. It was a great place to work on afternoons and weekends to make some extra dough, and in many ways, I loved the work.

When you get paid hourly, you have two choices—you can work the hours, or you can make the hours work. In other words, on slow days, you can either be lazy and pass the time doing as little as possible, or make yourself useful and find something to do.

Being the good employee I was, I wanted to make sure my employers got what they paid for, so I always tried to add value to my time (okay, almost always). Let me tell you—it's incredible what kind of work you can create in a coffee shop when it's slow. Scrubbing panels, restocking, reorganizing the attic, or cleaning up seeds and crumbs from nooks and crannies. I never felt good just standing around wasting my time. That's not what I was paid for, so I gladly dove into those different projects. I would highly recommend anyone being paid hourly to treat your work this way. You don't have to kill yourself trying to prove your worth, but find ways to make those hours count—as much for the employer as for yourself. When you bring value and purpose to your work, the hours pass much faster, plus you never know what you can learn by extending yourself to do a bit more. It speaks to the kind of person you are. By always giving a little more than what you're asked, not just at work, you will continuously find yourself on new ground. You become a doer and explorer, and you never know where that mindset will bring you.

For me, it brought me to the same place for ten years. Ten years! I tried to quit many times, but I had developed such a strong relationship with the quaint old place, and they simply wouldn't let me go, so for ten years I treated that coffee shop like my second home. If I didn't work, I went there to study, and even if I was only there on a *fika* (what we Swedes call a coffee with a sweet in the afternoon—preferably right at 3 p.m.), I couldn't help myself. If I saw a table that needed to be bused, I'd probably do it. I was working even when I wasn't working, and it was hard to shut it off.

I owe a lot to that coffee shop, and I'm pretty sure I wouldn't be where I am today without it. Not just because it allowed me to make money and save up for my trip to America, but because I learned things about myself that enabled me to take scary leaps of faith. I built up trust in myself, and that's not to be taken lightly, so if you're a teen reading this and considering getting a job, let me say—do it. You might hate it at times (lord knows I did), but it's a gift that will keep giving, in ways you can't even know.

However—and I'm getting to my point here—with all the things the coffee shop job gave me, the ability to be busy isn't serving me so well anymore.

Making yourself useful is one thing. Constantly being haunted by the idea you are useless if you are not being productive all the time can be damaging in so many ways. I've spent the last few years trying to unlearn this behavior, and it's been difficult. I had learned to attach my self-worth to how much I *did*. That's how I thought I showed my worth. But let's face it; when you've reorganized the cookie tray for the third time, you're not really adding value; you're just staying busy.

I understand not everyone relates to this feeling. (I'm married to a champion procrastinator, so I know this as a fact.) But I also know I'm not alone in being addicted to being busy. When you practice it for long enough and keep reenforcing the idea that your worth is directly tied to how useful you are, it becomes like a drug you can't shake.

On my journey as a climate optimist, the thing I keep learning is that I need to do *less*, not more. That in *doing less*, I'm actually *gaining more*, and in gaining more, I need less stuff.

Isn't that what we do? We're like trained soldiers who believe if we don't have schedules packed to the brim, we're not "good, worthy people," and if we're not doing things all the time, then our life is probably pretty bleak in comparison to other people's lives.

All you have to do is go on Instagram to learn everyone else is doing fabulous things, which makes you think you should be doing something fabulous too. But are these fabulous things actually bringing more value to our lives, or are they just filling a void and giving us something we can show to the world to prove we're worthy?

Don't get me wrong—brunches, meetups, and TikTok dancing are all wonderful. Bring more movement and real-life interactions to the people, please. But I challenge you—as I'm constantly challenging myself—to reflect on how often you're being busy just to be busy. How much of your self-worth are you attributing to what you can do and get done? If you take an hour of your day (or less, maybe even fifteen minutes) to just sit and do nothing, will you start thinking you're doing something wrong? And if so, can you please—for yourself and for the world—take a deep breath and say to yourself, "I see what you're trying to say, but actually, I'm doing the right thing, so let me just be."

Let Your Spirit Catch Up

Doing nothing is a beautiful thing. Doing nothing allows your spirit to land in your body and for your body and spirit to tune back into the world. I read a beautiful story of African porters working with some missionaries. The porters traveled so fast for a long distance that they had to pause. When the missionaries asked why they were pausing, they said it was because they had gone so fast they needed to let their souls catch up.[22] How beautiful is that?

How often do we allow ourselves to do this? We get rewarded for speed, hard work, and success, but very rarely do we hear someone say, "I can see you're spiritually caught up; that's so great."

I confess I am a notoriously bad soul-catcher myself, but I'm working on getting better, and the more I practice, the more value I feel I bring to the world. In those moments of stillness, I catch creative ideas for a new project. In those moments of stillness, I can clearly see my next move. In those moments of stillness, I understand what my life is about, and it becomes clear I don't need more stuff, food, or activities to make life meaningful; sitting here doing nothing but letting my soul catch up

---

22 https://www.paulborthwick.com/take-time-to-let-your-soul-catch-up/

is pretty meaningful in itself.

Don't get me wrong; it's good to be busy too. For the work in front of us building a better and more sustainable world, we need to accelerate and start sprinting. But in this sprint, in this climb to reach new heights, it's absolutely essential we pause from time to time to let our souls catch up. Because if we don't, how will we know we're sprinting in the right direction? How will we know we're embarking on the right climb? And most importantly, how will we enjoy journey there if we're not spiritually in our bodies, enjoying all the precious moments we were brought to Earth to have?

CHAPTER 46
Choosing Light

Sometimes it feels like it's not socially okay to be okay. It's almost like it's morally wrong to be too happy. You catch yourself feeling good about the world, then instantly remember the pandemic, climate change, and that so many people and animals are suffering, so maybe things aren't so good after all. And how can we choose to be a light when this is what we know? How do we find hope in a world that is so hopeless?

Because, my friend, it's the only thing we've got.

On days when awareness overload creeps in and I'm unsure what to do, thinking maybe I should give in and just accept the world is falling apart, to keep fighting, I remind myself I am happy. I tell myself at the end of the day, the light inside me is all I have, and I remind myself no one else is going to light it for me.

Next, I remind myself that if I choose to be a light, I allow others to activate their light as well. I think of one of my favorite quotes and remember light will *always* lead to more light for all.

"A candle loses nothing by lighting another candle."
— *James Keller*

When the guilt, worry, or fear kicks in, I remember I have the power to choose to be a light, and by activating that power, I become a candle in a dark room. With my own candle burning, I can help others light theirs too, and soon there will be light where there was nothing but darkness. Maybe making this choice won't automatically make things better, but it is a whole lot easier to see the dust on the floor and start cleaning up.

We long to be able to help others. We crave the power to change the world. Too easily we forget this power lies within, in our hearts, waiting to be unleashed.

Lori Ladd is a YouTuber who talks about the evolution of human consciousness. A message she shared earlier this year resonated with me a lot. She said we humans are living inside an arena, but we don't know it because we can't see the bleachers encircling us, holding us in place, so we barely know we're "stuck" here in the first place.

The shadows lurking in those bleachers are jealousy, greed, fear, and other cosmically imprinted pain—they are the shadows of societies past that created the rules we live our lives by. No one questions these "rules" because we don't consciously know they're there, which is why social injustices, patriarchy, and capitalism, among other things, continue to rule, subjugate, and oppress.

But all we have to do is shine some light on those bleachers, and voilà, there they are, our outdated fears and petty jealousies, our instinctual pain and traditional, self-inflected limitations, busted and ready to be questioned. Some of us are already there. We can see the dark forces holding us back, and we are so ready to a take big, bold step into a different kind of world. We long for a time to be free and able to cocreate something so much better, something so much more peaceful and beautiful than the world we know today. We know it's possible, and we know we can achieve it, as long as we become aware of those shadows lurking all around.

That is what dismantling broken systems looks like. That is what removing injustice and restructuring fragmented ideologies looks like. It looks like shedding light on the shadows and telling ourselves this is no longer okay. We're done playing by these stupid rules, and we demand something better.

This work is already ongoing, and we must remember this sort of dismantling can create some chaos and change. Old systems *are* crumbling, but we've yet to create something new, so hold on tight and stick with it—better times are coming. And while we're still in this arena together, we have one mission, to continue to be the brightest light we can be. By doing so, we invite others to partake in the light, and together we can illuminate the arena around us.

What's important to understand, which Lori Ladd said in the video, is you can't force someone else to see the bleachers. The only thing you can do is activate *your* light and help them see it for themselves.

This is what I remind myself of when it's hard. I close my eyes and tell myself I can't control all the dark forces in the world, but I can control my own light. I remember choosing to be light is an act of heroism, and helping light someone else's candle is a noble act that can have a ripple effect in ways we can't even imagine.

Our light is not a scarce resource we need to guard so no one else steals it. But our light is abundant. If we light a candle with another candle, the first candle still shines just as brightly. If we turn on another lamp in the room, the first lamp doesn't get dimmer. They both help brighten the room.

That's who you are in the world—another source of light that can help make the world brighter for everyone. You carry this light inside you, and you can choose to switch it on any time. I know it can be hard, and in some particularly dark places, finding the switch takes a lot of fumbling in the dark. But find it. Keep looking. Make it your mission to ensure that no matter what, there will be light again, and you won't stop until you find it.

That light is all we have, and it's amazing what we can accomplish when we turn it on and the arena becomes a place of awakening, hope, and excitement.

CHAPTER 47
AT's Tips for Climate Healing

Understanding how important my mental health is, especially as I continue in this climate work, I've established a few healing routines. I find when I weave healing routines into my schedule, I am more resilient, and the need for deeper work is rare.

You can try one, two, or all of them if you'd like. Of course, only try what you are comfortable with and only continue the ones that work for you. Just remember that finding ways to tune in and ground yourself, preferably daily, is incredibly important. You are an Earth Warrior; now start treating yourself like one.

Nature Dates

One of my favorite things is going on "nature dates." I don't do this every day—and it's harder in the winter—but I do try to squeeze them in whenever I can. As the name implies, you simply take yourself on a date with nature; you find a sacred moment when it's only you and nature. You can take a walk in the woods or on the beach, take a silent moment by a river, or my favorite, contemplate on a mountaintop after a good, sweaty hike.

A nature date is a little bit different than just a walk. When you're on a nature date, you allow yourself to tune in. You let go of the outside world and just exist for a while, which is a beautiful way to tap into yourself, your body and mind, *and* with Mother Earth.

I like to bring a snack—a salad or a thermos with coffee and some nuts and crackers. It makes the date a little more special. I try not to listen to a podcast or call a friend because that would take my focus away from the moment. I do, however, like to bring a journal just in case some thoughts come through that I'd like to jot down.

I call it a nature date because it makes it feel special and reminds me to focus on my date, Mother Nature. However, she exists everywhere, so you don't have to be out in the wild to do this. If you have a park nearby, that's a great choice, or maybe a river where you can go and sit for a moment. Sometimes, I plop myself at the roots of a tree in Central Park, which becomes nature dating at its finest. You can do this anywhere, even New York City.

If for whatever reason you feel nerdy doing this, don't worry. No one needs to know you're on a date with nature. Keep it to yourself and create a secret, magical bond only you and she know exists.

Walks

Since I can't go on nature dates all the time, I go on walks. In my opinion, this is the second-best when it comes to tuning in to yourself and the world around you. I go for walks multiple times a day, one first thing in the morning, one again around lunch if I can, and then another long one in the late afternoon. Some nights I also pop out for a short stroll before bedtime—it helps you sleep.

I'm a walker. This is the form of exercise I do. If you're into running, biking, yoga, tennis, CrossFit, or whatever it might be, you do you. Any kind of exercise is great for keeping your endorphins high and your anxiety in check. However, I praise walking because it allows you to tune in to yourself more than most other forms of exercise. Even if it's just fifteen minutes in the morning or at lunch, try to get yourself out for a walk. Be quiet if you can so you give yourself a proper break and let yourself *feel* nature. Feel the wind on your face, the sun on your skin, or the raindrops making the air so fresh on a rainy day. We have a saying in Sweden that I preach to almost everyone I meet: "There's no such thing as bad weather, only bad clothing." No excuses. Get yourself out there.

Hug a Tree

Naturally, I'm a tree hugger, and not just a figurative one—I actually love to hug trees. And you should too. I have a whole chapter dedicated to this later, so I won't go into too much detail about it here, but for me hugging a tree is a card you can always play to feel a little better. On days when I'm anxious, it can truly calm me down. But it's not enough to quickly walk up to a tree in passing and give it a little squeeze. You have to plant your feet firmly on the ground, lean your full body against the tree trunk, and then wrap your arms around it with love and sincerity. Rest your cheek against the bark and stay there for a little while. Let the energy of the tree transfer into your body as you feel the energy from the earth beneath you come up through your feet. Slowly.... Calmly....

It works every time, and the tree likes it too.

Hugging a tree fills my need for grounding and soothing, and I can hug a tree at any time. I try getting one hug in at least every day to keep that connection with the earth strong, and also to remind the trees I'm still here, still loving and supporting. If you think a full-on hug is a bit too much (especially if you live in a populated area), simply walk up and place your hands on a tree for a few seconds. A simple touch can be enough to fuel you for the day.

Music

What would we do if we didn't have music? I don't always listen to it on my walks. (Sometimes I like quiet time with nature; sometimes I love a good podcast.) But few things can bring your mood up like some good music on a power walk.

I find it important to allow myself to let the world and all its seriousness go for a while, a moment in my day when I can just exist with myself and the music. I allow myself to dream of good things to come and believe in the unbelievable because that's what music can do for you—it makes you believe in magic. I also love dancing in the apartment or having some good grooves on while cooking. Music can make any moment better.

Community

Now, I can't emphasize this one enough. Community is everything! I used to think I was alone in trying to save the Earth, but as I connected with like-minded people over the years, I've felt my faith in humanity grow and my anxiety continue to ease. Yes, we are still facing insurmountable problems, and I'm still not sure if we are going to fix this in time, but staying connected to people who are passionate, creative, driven, and determined to make a difference keeps me from losing hope.

A community can take many forms. You can join a local grassroots organization or join a bigger one, like Greenpeace or the Sierra Club. You can also form your own community among friends and find ways to stay engaged in the things you care about. Often, a community is simply an extended family where you can seek support and share a laugh, or a cry on those more difficult days, and it gives you a platform for learning and sharing information. You can find this extended family close by or far away. With the internet at our fingertips, you can connect with people all around the world.

A community can also be friends, neighbors, and/or your network on social media. Those who make you feel seen, supported, inspired, motivated, and loved are your community.

You can have many communities at once, serving different purposes, but never underestimate the importance of having people and energies around that help you remain in the light.

If we didn't have each other, we wouldn't have anything at all.

Good Food

For me, nutrition is everything. As Hippocrates said, "Food is thy medicine, and medicine is thy food." I truly believe and live by these words. I believe it applies to all aspects of our health, from physical to mental to spiritual. What we put in our bodies is so important, and although I sway from my regular diet from time to time, I do stay quite strictly with what I know my body wants. Right now, it happens to be about 95 percent plant-based with an occasional egg or some dairy here and there. I also avoid refined sugars, wheat, tobacco, and alcohol as best I can.

I'm not saying you have to follow these guidelines or give up on the things you love, but to me, a healthy and wholesome diet is essential.

I can tell after only a day or two if I've been diverging from my regular foods, and although these other foods seem like a treat at the moment, they never serve me for the better. I simply love my body too much to "junk it," and I also love these good foods too much to give them up. When it comes to anxiety and mental health, eating well is everything. As soon as I'm off balance nutritionally, I can feel it undermining my mental health. The two are interlinked, so I notoriously practice eating good foods—and with joy.

Positive News

As mentioned previously, finding ways to infuse positive news into my life is critical if I am to continue steeping myself in all this climate awareness. I need to know we're headed in the right direction, and despite the apparent lack of hope, the movement forward can still grow strong.

Each time an especially depressing report or some really bad news reaches me, I need to find some light at the other end of the tunnel. I need to keep that light shining so I know I still have reasons to show up, and I seek light with determination and passion because I always find it.

Meditation

Truth be told, I'm not the best meditator. I do meditate from time to time, but I have yet to establish a regular practice. With that said, meditation is a proven life-changer in many ways, and it has significant positive effects on your brain, wellbeing, and mood. If you meditate, amazing—keep going. If you've never tried it, I highly recommend giving it a shot.

Don't expect unbelievable results right away, and be easy on yourself. Just go "sit" for fifteen minutes and let your thoughts wander on their own. Don't get frustrated if you become bombarded with thoughts all of a sudden—this happens to the best of us. The art of meditation is not having no thoughts at all, but being able to observe them as they come without paying them any attention. Think of them as traffic driving by. You can watch the cars go by without interacting with their drivers. Do the same with your thoughts. Observe, let go, and give space.

Journaling

Finally, I'll mention journaling because it has helped keep me sane for many years. We've already looked at some journaling guides, which I recommend trying if you're curious. However, journaling does not have to have a purpose or follow a prescribed prompt. More often than not, I write just to understand what's going on.

I can feel lost or out-of-body, and when I do, journaling is a powerful tool. If I find a moment to sit down and put pen to paper, the words soon start to pop onto that paper, and things begin to make sense. One sentence after the other, I put words to how I feel, and suddenly I see clearly what's actually going on. It's like I give my inner self a chance to have a word, and what she has to say is always so wise and intelligent.

You don't have to journal to figure something out. Sometimes you simply want to express how you're feeling or hurting. You also don't have to be in a bad place to journal. Some of my favorite journaling sessions are when I get to brag to my higher self about all the good things going on and record on paper how amazing I feel in that moment.

If journaling works for you, hooray—it's perhaps the simplest and cheapest kind of therapy there is.

PART FOUR
CHOOSING OPTIMISM

CHAPTER 48
Optimism Is a Tricky Thing

The following pages will be dedicated to the science of optimism and the actual work that goes into being optimistic. My hope is whether you're a hopeless optimist or hardcore pessimist, you will have a different view on optimism after reading the next few chapters because what I've learned is optimism is a tricky thing. You can't just choose to be optimistic. If that's all you do, it won't last long. You can say you are an optimist without knowing why.

Hope can exist without proof—it's the last thing we let go of when everything else seems to have left us.

Optimism, however, doesn't work quite the same way. If you intend to stay optimistic for a long time (your whole life maybe), you have to understand the work that goes into being an optimist. Having spent the greater part of the last decade trying to figure out how to be optimistic, and a climate optimist to boot, I hope I have some valuable insights to share.

Sometimes optimism feels like a sin. Have you ever felt that way? You feel like you're not allowed to be optimistic because everything around you seems to be going to shit. When people are hurting, animals are dying, and signs of climate change are everywhere, who are you to be optimistic? When injustice continues to show its ugly face and society appears to be going backward, who are you to keep a smile on your face?

Maybe you've never felt this way, but I sure have. I start out having a really good day, and then someone tells me something, and suddenly, I feel like my smile is inappropriate. Maybe I should get my head out of the clouds and swim in the river of ugliness and truth for a little. Maybe then I'd feel like a better citizen—one who actually cares.

I think it's normal to feel this way, especially considering the world is filled with so much negative news. Someone telling you it's okay, good even, to be optimistic can be like having someone say it's okay to have a crush even if the object of said crush doesn't like you back. You get permission to—no matter the outcome, no matter the odds—feel how you feel.

It's a nice feeling, isn't it? But optimism doesn't ring true for everyone. I've grown to understand that optimism is a challenging word to understand, one with many different meanings to different people. Let's see which one resonates best with you.

For some, optimism has been drained of its actual meaning. Just like "miracle" or "love," it can be overused to the point where it has lost its meaning. If everything is a miracle, maybe nothing is a miracle—it all becomes the same. And if a thirteen-year-old texts "I love you" to people they barely know, maybe those words mean something different. When they tell you they love you, do they mean it? If they can "love" someone they barely know, what does love even mean to them?

Optimism can also be triggering. For someone who's been through or who might currently be dealing with a lot of pain, hearing someone say you must remain optimistic can be met with negative reactions. Either you feel like the other person can't relate (which might be the case), or you feel even more alone, like the rest of the world doesn't get it. Sure, you'd love to be optimistic—it seems great—but how can you be? How can you see the light at the end of the tunnel when everything in *your* world is so dark?

I believe the biggest misunderstanding with optimism is that it's something we get to choose or not choose. We ask one another, "Are you an optimist?" as if it's something we get to choose daily, just like settling for an outfit with a matching belt. "Should I go black or brown with these shoes, honey?"

In many ways, we choose optimism the same way. We open the doors to our emotional closet and say, "Okay, I'm going to wear some optimism today. I think it'll suit me well." The tricky thing with this approach is we rely on the outside world to provide the optimism. Just like we rely on stores to sell the clothes we need, we depend on news articles and social media posts to fill our emotional needs. In other words, we need signs of good news we can choose from.

If you've ever turned on the news, you know this can be an uphill battle, especially if you're looking for climate-positive news. Trying to be a climate optimist in a world with nothing but news of climate catastrophe is very much like playing golf with a ski pole in the dark. Frustration grows, and in that frustration, a feeling of futility grows. Hopelessness. Maybe you're not a golfer after all. Maybe you should give up on our climate altogether....

I was there for many years. In my early days as a climate optimist, I thought if I could only focus on the positive, ignore the bad, and continue wearing the smile I was given, I could show myself and the world that there is still hope. I could be the sun in all the clouds and help other people see it too. But I was living in denial, and the interesting thing with any sort of denial is that although *you* are not actively paying attention to whatever you're trying to ignore, your body still is. As you're consciously focusing on positivity and believing in a better world, your body and soul soak up all the negative and store it for a better day. One day, they hope, you'll take time to actually deal with it, so like a sponge in the rain, you get heavier and heavier as the years go by.

In failing to be a (true) climate optimist for quite some time, I learned something. I learned that when you choose optimism and pair it with denial, something interesting happens. Your consciousness looks at the bright side while your body remembers everything else. While you're out there working hard on seeing only the light, your heart, soul, and body remember the pain. And while you're not paying attention, that pain builds up inside and slowly starts to take over.

It took a few years of being a climate optimist to fully understand what real optimism is. I know now you must be aware of the negative and see it for all it is. You must know your heart can take it, and from that acceptance, you can begin to heal. Because when you are absolutely aware, you don't fool yourself into optimism—you *create* it. You ask yourself, "Knowing all I know now, what can *I* do to make it better?"

And in doing what you can, in being the change you wish to see in the world, you become your own source of optimism.

I call it being an optimist in action (OIA).

To be an *optimist in action* is to be the change you wish to see in the world, and in doing so, fuel your own optimism from within. You do it for the world, yes, but more than anything, you do it for yourself. Because not only will it fuel happiness hormones in your body (more on that later), but you will also begin to see the world with whole new eyes.

You learn it's okay to be wrong because only when you're wrong can you learn something new and do better. And as you leave room for failure, you also leave room for improvement, and that's where the innovation sparks.

As an optimist in action, you're not just hoping things will get better, you're proving to yourself they can. Even if it's just small things like saying no to a plastic lid at the coffee shop or taking the bike instead of the car, you begin to exercise this optimism regularly. Small actions build character and set the framework for who you are, which will completely change how you view the world and what you believe is possible. Being an optimist in action, you start to believe in *change.*

Remember you can't just choose optimism. If you do, that optimism will soon fade, and it'll be impossible to escape the negative energies around you. To *remain* an optimist, you must find ways to *create* optimism—but that is also where the fun begins.

The more optimism you create, the better you feel, and the easier it'll be to create more. Relying on sources beyond your control for reasons to believe in a better world is silly at best. In many cases, this wishful thinking can lead to despair, inaction, and sometimes even anxiety and depression. That is not what we need more of right now.

The world needs people moving forward and asking questions. The world needs people who are willing to show up to do whatever they can to bring a better world closer. For that, we need to keep this *one* truth in our hearts—optimism fuels action.

Not quite sold on optimism yet? Don't worry. I'm about to get really nerdy and tell you all about it.

CHAPTER 49
Understanding Optimism

I've received some hate for my optimism. I will never forget when my ex-boyfriend yelled at me to please stop being so damn optimistic. I just stood there, silent and stunned. Of all the things I may have expected from our fight, I didn't expect that.

What hit me was that, apparently, optimism can annoy people. Others think I'm not being realistic. These folks usually claim to be realists while pointing out my naivete. Well, I'm about to eyeroll them back because, get this, *optimism is hella important*!

It is saving the world important, and I have proof.

Not only is optimism important if we want to change the world, but it's also been actually scientifically proven that optimism plays a vital role in our health, wealth, and happiness.

I think we can agree that elephants don't fly, so it could appear *realistic* to say we won't see one outside our window tomorrow. But remember this—humans didn't use to fly either, and anyone who claimed humans would fly in the future was probably ridiculed relentlessly. Only now humans do fly, all thanks to two optimistically driven brothers who just wouldn't give up. I guess the joke was on the realists.

Okay, radical optimism and toxic positivity are very different. I'll admit, I *may* have leaned toward the latter during periods of my life. But I've also come to understand real optimism is earned and you actually have to work for it. As you already know, I call it being an *optimist in action*, and it basically means if you show up for the work, you're living the change, and the good thoughts and feelings will come through to reward you. It's like optimism on autopilot—you need it to get going, but the work itself creates more.

A lot of realism is infused in that way of being. You don't just hope for things to happen; you understand that only when you apply realistic measures and strengths to an issue will you actually be able to reap a reward. However, as you go on, your view on that realistic outcome will probably change, as you become a living example that we can do hard things and push boundaries for the better.

The question shouldn't be whether you see yourself as an optimist or a realist—the classic "Is your glass half empty or half full?" Instead, reflect on where your focus is and what you actually want to have happen. Being aware and choosing to be the change you wish to see, you become an optimist—a realistic one. What you should really ask yourself is, "Do I want to be realistically optimistic or would I rather stay complacent in my beliefs?"

To give you a hint: one keeps you stuck, the other gets you moving, so choose consciously.

I've intuitively known my whole life that optimism is good for me. When I was just a little girl, I would write myself happy notes and hang them on my door before going to bed. That way I knew I would wake up with an encouraging message to start the day. I did this way before I learned anything about the power of thought or the law of attraction. I simply recognized that when I made myself feel a certain way the first thing in the morning, my chances of having a good day rose exponentially.

I later learned that this is called "priming" and has been scientifically proven to have wonderful effects on people's happiness and careers.[23] You're basically tuning yourself in for success like you would tune in a radio station to capture your favorite music. And it works—thinking that your day will be good significantly raises your chances of having a good day.

I will explain this and much more in the coming chapters.

Before we get to that, it's worth pointing out that optimism is a developed skill only humans have. Thanks to our frontal cortex (the most recently developed part of our brain, the part right behind your forehead), we can do some pretty awesome things. For one, we can think of something that has yet to happen—we can *imagine* a reality yet to be.

---

23 Williams, Lawrence E. and John A. Bargh. "Experiencing Physical Warmth Promotes Interpersonal Warmth." Science. 322.5901. October 24, 2008. https://www.science.org/doi/10.1126/science.1162548

For example, you can think about what you will eat for dinner tonight or what you are going to wear to the party on Friday. In other words, you can *envision* things that are not right in front of you right now. This is pretty cool stuff. No other species can do this.[24]

The other awesome thing is you can create an *expectation* of what that yet-to-be scene will look like. For example, you can expect your veggie curry to be delicious and your outfit on Friday night to make you look amazing. You can also dream up a scenario where your outfit will help you catch the eye of the person you have a crush on; they will ask you to dance, you'll kiss on the dance floor at the end of the night, fall madly in love, and eventually get married and have a house, dogs, and kids, while staying together forever. (Let's be real—everyone has been here at least once.)

Have you ever hoped for something to happen—even believed it would—only to be disappointed? They never proposed, the trip was ruined by bad weather, or your boss decided to promote someone else. Whatever it is, I'm sure you have many stories because the truth is we all do—we all dream of things that never happen. The interesting fact is most of us are much more optimistic than we're aware. In fact, 80 percent of the human population is inherently optimistic, walking around believing things will be better than they actually turn out to be. (This goes for pessimist too.)

Hang on a second—I'm living in an illusion of a world that probably will never be?

That's right. Studies show the majority walk around with an optimistically biased mindset, simply called the Optimism Bias.[25] We simply envision future scenarios as better than they actually turn out to be. I understand it might sound incredibly stupid to fool ourselves all the time, but doesn't it make you wonder *why* this is? How is it that our brains, the incredible computer we have inside our heads, is wired to think of future outcomes as more favorable than reality will allow? Is there maybe something important about our ability to think this way?

Absolutely, and soon enough we will understand how incredibly important this talent is. Our species is not dumb, and in many ways, op-

---

[24] New studies on marine mammals are starting to show they, too, might be able to plan for the future, so let's not think we're so special we close our hearts to them.

[25] Learn more about this topic by reading *The Optimism Bias: A Tour of the Irrationally Positive Brain* by Tali Sharot.

timism isn't just good for us; we need it to survive. Sure, hoping your hottie will kiss you at the dance isn't life or death, but the very fact that we can muster hope is what has taken us through millennia of change and brought the human species to where we are today.

If you don't think there's life and water on the other side of those sand dunes, how would you muster the strength and willpower to trek through the scorching hot desert? If you don't think your invention will enhance the lives of millions and make you some money in the process, how could you justify all those hours in the lab?

You could say human achievements throughout time—athletic, societal, social justice, scientific—are the result of grit, sweat, and tears, and maybe a bit of crazy too. That's all true, but none of the grit, sweat, or tears (or even the crazy) would be rational in the first place had there not been a vision to fuel it, and that is what optimism is.

As you can probably tell, I'm a sucker for all this, but I don't want to be alone in this optimist squad, so in the next chapter, I will share my three main reasons for becoming a fan of optimism.

Welcome to the world of positive thinking.

CHAPTER 50
Three Reasons for Optimism

During a recent solo trip to El Salvador, I took my first surfing class. It was a childhood dream of mine that I finally got to realize. But before they walked me into the waves, I needed to go through a necessary training session. I was shown how to position my legs, how to pump myself up with my hands under my chest, and told how important it was "not to lose a second." Once the wave was beneath me, I had to act fast. But most important, I was instructed, was that believed I could do it. Without that, the instructor said, it is impossible to surf.

I've tried new and scary things before, so I knew exactly what he meant. I went back to the triumphant moment when, at twenty, I decided to learn how to snowboard. I had taken the ski lift to the top of the mountain first thing in the morning and snowboarded all the way down.

I knew which mindset to bring to surfing.

On that day ten years earlier, I had wanted to give up snowboarding. After two days of trying and failing, and with a bruised body that hurt no matter how I turned, I felt defeated. It was no fun anymore, and it felt like I would never learn. I had started too late; I was too old, I told myself. Besides, my brother, who had promised to teach me, was starting to get sick of me falling, so with a "Sorry, Sis," he left me on the slope and disappeared.

I almost gave up, but then I remembered something I had learned about the power of the mind and how powerful it can be when we focus correctly. That night when I went to bed, I closed my eyes and visualized myself snowboarding. I saw how I would take the lift up as soon as the ski hill opened, get off without trouble, buckle on my board, and then gracefully glide down the mountain. I hadn't had one successful ride yet, but I saw and *felt* it in my vision, and I knew I could do it. I closed my eyes and drifted off to sleep.

Had it not been for that priming moment the night before, I'm pretty sure I would still be defeated and disappointed on the snowboarding front. Grit and trust in my visualization got me down that hill, and for that, I am still proud. I haven't stood on a snowboard since, but the triumph lives within me to this day.

I'm sure you have your own story, times when you wanted to give up, when it seemed impossible to succeed, but a voice inside you told you to keep trying. What pulled you out of your frustration and made you try again? What was your incentive for not quitting?

It was the belief that you could do it. It was an inner knowing that if you kept trying, you would succeed. That, my friend, is optimism—it's a thought you create that tells you a desirable outcome is possible so it's not yet time to give up. Think back to any moment when you made a real change and you will find optimism played a huge role in your success.

The truth is we need optimism. Without it, we simply wouldn't do much. Our brain developed the frontal cortex, the part that enables us to imagine realities that are yet to be, because it *moves us forward*. Thanks to this nifty detail in our brain, we've come a long way as a species.

Every innovation and every forward-thinking movement was birthed in this part of our brain, nurtured in the space where we can *dream* and envision new things. But the ability to think of these desirable outcomes is not enough; we also need to *believe* they can actually happen.

Neuroscience confirms this theory. In *The Influential Mind*, Tali Sharot writes about how our brains are programmed to associate "forward action" with reward and pleasure. We move *toward* something because we believe it is going to serve our interests. We seek to achieve things because we want the outcome achieving them brings.

Simultaneously, our brain is also built to trigger a "no go" response when faced with the possibility of loss. In other words, when we don't want something to happen, we're programmed to freeze. This may explain why when you're overwhelmed, the stress itself makes you incapable of acting. You want to act, but somehow you can't seem to get yourself to take a single step.

Coming from a background in marketing, I am well aware of this. Marketing is basically learning how to play with people's brains to get them to do or buy whatever it is you're promoting. Advertisers know if you want people to act, you need to play on positive emotions. This is why you don't see a commercial telling you how *awful* you'll smell if you *don't* buy their product; you see a beautiful, smiling woman or man showcasing how *wonderful* life is once you've bought it.

Say what you want about advertising, what we should learn from this is that positive feedback drives action much more effectively than warnings or threats. This is worth noting for anyone trying to spark action in the world. (We'll go into detail about this in the next chapter.)

Here are my three favorite reasons optimism should play an essential part in our work, communities, and lives.

> *"Perpetual optimism is a force multiplier."*
> — Colin Powell

ONE—You Need It to Act

This part is fairly straightforward. If you didn't think something is actually possible, why would you even try? If you didn't *believe* we have a chance at cocreating a better world for the future, why would you even consider taking action? It's as simple as that—our brain runs on optimism to get things done. It needs to *believe* something is A) possible and B) favorable to find the motivation to act.

Sure, optimism bias leaves us thinking things will be better than they usually turn out, which might seem like a downer, but the key here is not to be disappointed when things *don't* turn out the way you expected. My dear friend Mo Gawdat, the author of many books including *Solve for Happy*, defines happiness as how good you are at dealing with missed expectations. How well can you manage and *alter* your expectations if things don't go as planned?

If your plane ends up being delayed, how do you react? If your blind date isn't as fabulous as you expected, how will you deal with it?

The secret to staying optimistic, driven, *and* happy is daring to believe in the crazy things, finding the courage to go for it, and then being able to pivot when something doesn't go as planned. That is what being a true *optimist in action* means. You pick up new hope when the old hope runs dry. You grow new optimism where your old optimism meets dead ends. You need optimism to act—again and again and again.

Recap for Optimism for Action:

- The probability of loss triggers inaction, a no-go response in our brains, when faced with negative information.
- On the contrary, we associate a possible reward with action, triggering a "let's do it" response when we believe in something good.
- We're optimistically biased because we need this frame of mind to think creatively, boldly, and strategically. Without optimism, nothing would ever get done.

TWO—Enhances Your Chances

> *"Optimists are 40 percent more likely than pessimists to get a work promotion within the next year."*
> — Time Magazine

If you care about making money, you might want to give optimism a try. When the numbers are this astounding (see quote above), it amazes me no one teaches this in school. When our chances of success are significantly higher with an optimistic mindset, why isn't optimism a mandatory part of everyone's coursework? (I have studied business and marketing in both Sweden and America, and I have yet to come across any courses of this kind, but please contact me if your experience is different.)

What we have to understand is that enhancing our chances with optimism has nothing to do with wishful thinking. Although I'm a firm believer in some sort of magic, it doesn't have to do with the law of attraction either. The reason optimism enhances our chances of earning money, happiness, and success is because our internal alert system is heightened, meaning we will process information differently and make very different choices if we believe we have a chance.

A study by neuroscientist Sara Bengtsson, in which she manipulated positive and negative expectations among students while their brains were scanned and tested, showed that priming the students with different words (smart, intelligent, and clever, or stupid and ignorant) led them to respond differently to mistakes they made. If they had been told they were smart, an alert was triggered inside the brain when they got a wrong answer, showing enhanced activity in the anterior medial part of the prefrontal cortex. This is a region where self-reflection and recollection take place.

For those fed negative words, no such activity was spotted, and they didn't reflect further on their mistakes. In other words—people who were primed to believe they were stupid received the news of a wrong answer as something to be expected. Meanwhile, the students who were told they were smart caught the error and recognized something must be wrong.[26]

This result tells us people who think highly of themselves are better at assessing information and growing through life. An optimistic person, when faced with adversity, will fire up the needed brain activity to learn, reflect, and grow. With this growth comes new ways of thinking and the ability to identify and find the courage to open new doors. That is why optimistic people are much more likely to get promoted—they simply go after the opportunities at hand.

Optimistic people are 40 percent more likely to get a promotion within the next year simply because they deserve it. Optimism allows them to process information in a different way and look at reality from a lens of lessons, challenges, and adventure. They will see silver linings when their pessimistic counterparts see nothing but dark clouds. They will look for those strings of hope and curiously give them a pull to see what they might lead to. Wishful thinking doesn't magically give them what they want. Hurdles activate their brains, triggering the action needed to take action, adjust strategies, and keep moving forward.

I get challenged on my climate optimism a lot. How can we afford to be optimistic with everything going on? I respond, how can we *not*? If we prime ourselves to believe the world is going under and time is running out, every news article will simply confirm our beliefs and keep us (mentally) stuck where we are. No activity will be sparked in the brain that makes us think, *Hang on a second. This is not what I want. This is not the outcome I'm hoping for.* Instead, the silence in our pre-frontal cortex will speak for itself and no new ideas or creative thinking will be set in motion.

Just the fact that having an optimistic mindset enables those little

---

26 To learn more about this topic, check out Tari Sharot's book *Optimism Bias*.

bells to ring whenever we approach an outcome we don't want says it all. We need those bells to ring in everyone's brain, and they need to ring often. We have to be able to spot unfavorable outcomes and find the courage (and curiosity) to reassess and find new ways. We have to aim for a climate justice promotion all across the globe, one that brings more abundance, clean water, and safe air to breathe for everyone. We need a better world, and we need it now, so let's bring on the optimism and act.

To home in on my point, if optimism enhances our chances by 40 percent, how is it *not* the most important thing we have? If we don't think a climate-just future and world is possible, how will we make the choices to manifest one? If four out of ten optimists are more likely to have success than the average doomsday-folk, then I'd make it my core mission to activate as many optimists as I can. Oh wait—I already have! ☺

THREE—Live a Better Life Now

"Even if that better future is often an illusion, optimism has clear benefits in the present. Hope keeps our minds at ease, lowers stress, and improves physical health."
— Time Magazine

Imagine you have your dream road trip planned for this summer. You and your best friends are finally going on the trip you've always dreamt of, and you wish you could fast forward to summer tomorrow. As the dreary days of winter crawl by, you daydream about open roads, breathtaking landscapes, quirky rest stops, and all those unpredictable out-of-the-blue adventures. You dream about fun spontaneous stops at local bars and roadside selfie moments to die for.

But summer is still months away, and ice-cold winds and damp air are currently holding a tight grip on you—and on your car. It's the kind of weather that can make anyone or anything cranky, metal and human bodies alike. If you want to use your old mate for the road trip this summer, you can't let winter get to it. You need to keep it in a garage whenever possible, check and change the oil regularly, and get it serviced when needed. If you don't, you run the risk of running into trouble when your trip is about to begin.

Think of yourself and life the same way. Hopes and dreams of something good to come will directly affect how you treat yourself and your body right now. This is what optimism does to our health and wellbeing—it gives you reason to take care of yourself so you can stick around for all the fun to come.

Saying optimistic people are happier and have a brighter outlook on life might be stating the obvious, but you might be surprised by how beneficial this frame of mind is. For example, did you know optimistic people are 50 percent more likely to reach the age of eighty-five than their pessimistic counterparts? A Harvard research study of 70,000 women over eight years showed the most optimistic were almost 30 percent less likely to die of major illnesses like cancer, heart disease, stroke, respiratory disease, and infection compared to the least optimistic participants.[27]

Turns out (and back to more things that make a lot of sense) optimistic people are inherently better at taking care of themselves.

If you believe there's something good to look forward to, you'll want to be around to experience that future. If you have a bright outlook on tomorrow, you will treat yourself with love and respect today. That is why optimistic people are more likely to take their vitamins and eat good foods, workout, and seek activities that tend to their mental health.[28] Optimistic people simply care for themselves better than pessimistic people, which directly translates into healthier and happier lives *now*—no matter the outcome of the future they're projecting.

The trick about optimism is you must be willing to stay open-minded and pivot. When things don't work out as planned, you must be willing to grow and move on. When I was applying for my very first apartment as a student back in Sweden, my hopes were up to the stars. I was so excited about finding the apartment of my dreams, I spent days daydreaming about how I would decorate every corner. I showed my dad a sketch of how I was thinking of furnishing the place, but the look he gave me was filled with caution.

"Are you sure you want to have such high hopes for this place? What

---

[27] Kim, Eric S. et al. "Optimism and Cause-Specific Mortality: A Prospective Cohort Study." *American Journal of Epidemiology*. 185.1 (2017): 21–29.

[28] "Optimism and your health." *Harvard Health Publishing*. May 1 2008. https://www.health.harvard.edu/heart-health/optimism-and-your-health

if you don't get it? You'll be so disappointed."

For a second, I contemplated what he'd said but briskly concluded it was worth it to dream big. If I didn't get it, it simply wasn't meant to be, and if imagining myself living there could somehow bend the universal rules in my favor, I would give it all I had.

I did get the apartment, and I loved that place so much, so for this one occasion, I guess my wishful thinking worked. However, sometimes—many times—things did not play out the way I wanted them to, and those are the times I've found myself growing. I learned it's how you deal with those missed expectations and how fast you tune your heart into a new station that determines how happy you truly are.

After twenty months of COVID, Arthur and I couldn't think of a better way to spend Christmas than in a new home. We made the bold decision to *buy* an apartment in New York, but did not know what a process that would turn out to be. The months dragged on as we waited for lawyers to finalize the paperwork, but we still hoped to get in before Christmas. I prayed, I worked with my magic, I envisioned us packing up our boxes and putting up the Swedish Christmas star in our new home—I even thought of how I would tiptoe out of the bedroom on Christmas morning and put on coffee in the kitchen while it was still dark. I planted all the magic I could possibly think of to ensure we'd be in before Christmas, and as the days crawled closer and closer to December, I brushed the doubts away and continued to hope.

We finally got called in for an interview with the co-op board—scheduled for December 21. The relief and excitement of finally reaching the final hurdle were mixed with a dose of disappointment, realizing there was no way we could be accepted, close the deal, and move in before December 24.

I gave myself a day to accept this new reality and wipe my mental vision board clean. Dealing with the emotions that might surface with missed expectations is essential for moving on, and as the news slowly settled in, I could go back to abundance and receiving mode. Maybe Christmas spent with family was more important after all, and maybe things were meant to be this way. With a clear mind and a new abundance target in sight, I was back to gratitude, trust, and magic—how exciting it was that we were buying an apartment in New York!

These stories serve as reminders of how well optimism presents itself in my life. Had I told myself in July when we first started looking for apartments we wouldn't be moved in by the new year, I might have collapsed in despair. At that point, I was so ready to settle down and have a place to call home. Not knowing the entire timeline helped me stay healthy, happy, and focused on the now. The optimism and the wish for a specific outcome brought light to the present, which enabled me to focus on all the other tasks. It was the fuel to my engine that kept me going, and I truly illustrated how optimism bias plays its part.

Think back to when a similar story played out for you. When did you wish for something but were disappointed? How good were you at dealing with those missed expectations? Did you pick yourself up to set new sails, or did you lose yourself to the waters completely? Also, can you remember the timeline leading up to that event? Try to recognize how hoping something would turn out in a specific way helped you stay the course. You will probably realize that optimism was the fuel to your fire all along, and if you hadn't *believed*, none of the other good things would've happened either.

As it turns out, we need optimism to find the motivation to act, but it also enhances our chances of getting what we want. Think of it as the electricity to your car. You need to charge your car to be able to take off, but you also need the same fuel to *stay* on the road and reach your destination. Optimism doesn't just rev up the engine; it significantly increases your chances of getting where you want to go.

> *"Despite the struggles that inevitably lie ahead, we must move forward with hope—not just for the sake of the next generation, but for our own sanity too."*
> — Susanna Schrobsdorff for Time Magazine

CHAPTER 51
Anger

Anger is an interesting thing. When you feel anger, it's like nothing in the world could stop you. It doesn't matter what someone else says; all you can think about is how angry you are, and if you're not careful, that anger can easily consume you. Having dealt with anger issues my whole life, I know this more than anyone. But my perspective on anger changed when I met an incredible guy named Mike Veny who taught me that having anger issues is actually a good thing. Or as he would put it, it's not your issue, it's your *superpower*.

Veny is a coach, author, and public speaker focused on mental health, with expertise on the subject of anger. Because he had also dealt with anger his whole life, he and I instantly clicked, and a deep friendship was soon established. I've learned a lot from Mike, but something he said really stood out for me: "When you learn to sit with your anger and *feel* it, instead of simply pushing it away, that anger will quickly evolve into passion, decisiveness, and successful leadership."

We have a lot of reasons for anger when it comes to climate change, and maybe that is a good thing. When you get angry, there's no room for bullshit, and you quickly find ways to get to action. Without anger, maybe we wouldn't know to act at all. However, knowing how to use anger for good and finding ways to let it *empower you* instead of consuming you is essential. That is what this chapter is about.

As further expressed by Mike, anger is an outward expression of power. You come to the realization that something needs to be done differently, that something is not right. When it comes to climate change, so much is not right, so much needs to be done differently, that we have many reasons to feel anger. When families have to flee because the weather has changed so much they can no longer grow crops, people in poor communities suffer from air pollution, heart disease, and asthma, and when homes around the world are wiped out by storms, our basic human needs are not being met. Climate change is so incredibly unfair it would bring tears to Mr. Scrooge himself if he was given a lesson.

To care about climate change is to care about humans, and when you realize greed and the need for power among the few keeps hurting the powerless, it can stir up a lot of anger. It's deeply rooted in our hearts to act up when things are this unfair. If you're young and furious, I understand you more than anyone. If I were a kid and learned about global warming, a manufactured crisis ruining your chances of a safe environment to grow old in, and I realized that people who knew about it haven't done shit, I'd be pretty angry too.

Anger is good. Without anger, nothing in the world would ever change. No justice would be had, and no progress would be made for people who deserve it. We need anger to spark the action that will get us out of this mess. However, anger can create a lot of trouble if that's all we hold on to because it blinds us from seeing anything else.

Similar to fear, anger creates a tunnel vision, and all you can focus on is the here and now. Rationality soon goes out the window, and you lose yourself to your fury. To actually manifest *change*, we need to know how to funnel the energy anger gives us into something truly productive.

Think of anger as the bouncer at a highly popular nightclub. It doesn't matter how many tricks you try to play, if the bouncer is told only pink unicorns are allowed to come in, then only pink unicorns will pass through the rope that night. Your brain is very much like a nightclub, and with an angry bouncer at the door, few other things will make it in—no rational thinking, no forgiveness, no visions for a better world. Only anger and the thing that fueled that anger in the first place will come in to party. That's tunnel vision for you.

Strong emotions like anger and fear will (for good, primitive reasons) block you from seeing anything other than the very thing that made you scared or angry in the first place. This is a fantastic survival mechanism. Back in the day, when faced with a hungry tiger, you didn't have time to think about trivial things like where to find your next meal or that the cut on your foot is starting to hurt. In that second—eye to eye with a tiger—

your brain says, "Forget about all that, and save yourself. Now!"

Since our brains haven't changed much since the Stone Age, we still have the same automatic response to things that scare us or make us angry.

Think about a time when you were really upset with a friend and try to remember where your mind was at in that moment. I bet it was hard to focus on homework or anything else. Critical thinking was tossed out the window. Perhaps you even considered doing something dumb, like texting mean words or telling lies, just to get revenge.

We do things we're not always proud of when the angry bouncer is blocking the entrance to our mind's door. It makes it hard to think critically about anything else. What we have to remember is the bouncer blocks all the good stuff too, like creativity, joy, optimism, and hope, so although fear and anger can be great for *activating* action, we need more to create actual change.

We can use fear to run away, and we can use anger to muster the courage to punch someone in the face. It's amazing the courage we find when we're angry enough. But for the bigger shifts to happen, we can't keep running and punching people forever. For that, we need to broaden and maximize the powers of our fascinating minds.

Therefore, we need to find ways to incorporate our anger so it can serve our greater purpose. By becoming one with our anger, it can, as Mike Veny so passionately says, become your superpower.

Usually, when we don't deal with our anger it takes over, and we *become* our anger. So, learn to find a way to not detach from it, but pause and connect with it instead. Feel it, ask it questions, and see what's really behind it. When you allow it to stay in your body and be expressed in certain ways, anger is possibly the best emotion you could ever have because it leads to action.

Anger leads to action. It leads to change, boundaries, protection, and decisiveness…it's putting your foot down! Think about kids. One of the first ways they know how to express their frustration is by stomping—they're literally putting their foot down. And what are they doing? They're not realizing it but they're actually taking their power back, and that is what becoming one with your anger is all about. You're taking your power back.

Here are some tips for incorporating your anger and using it for something good as taught by the brilliant Mike Veny. I recommend his practices because he actively seeks ways to channel his anger (before he goes on stage, he taps into his memory to find something he's angry about), and he's one of the happiest and most grounded people I've ever met.

As a climate optimist, you get to be angry, worried, *and* optimistic at the same time. In fact, learning how to embody all these feelings in a grounded and balanced way is how you will become a force to be reckoned with.

1. Lower body workouts, like boxing and squats. When you activate your lower body, you allow your anger to travel from your head and chest down into your legs. This takes the anger out of your head and allows you to stand on it. It becomes a powerful force to center you and plant your feet steadily on the ground. Boxing is a great anger exercise for two reasons. You get a chance to metaphorically punch the thing (or person) that makes you angry while also activating your lower body. But find a workout that works for you. It doesn't have to be boxing.

2. Meditate and/or journal on it. When you allow yourself to sit with your anger instead of acting impulsively, you give yourself a chance to get to the core of the matter while also landing in a much more grounded place. Your deeper truth comes through in those moments, and you can act from wisdom and love instead of from ego, which can make all the difference in how you end up handling the situation. Since we need both anger and visionary, open minds to tackle climate change, it's essential we do grounding work like this as much as possible.

3. Any kind of movement. When you feel anger take over, it's usually because it's getting to your head. Maybe you even start to feel like you're stuck in your own body, a prisoner who can't escape. Something as simple as a short walk, a little dancing in your living room, or jumping up and down in one place can help you shake this feeling. Animals will shake to get rid of fear. That way they can get out of the paralyzed state they're in and find the strength to run.

You can do the same with your anger, so shake it out. And maybe do some squats too!

"Once you find your way back to your anger, you will almost become dependent on it. I need my anger, now I have to go looking for it! When you learn to work with anger, it's incredible."
— *Mike Veny*

CHAPTER 52
How to Be an Optimist in Action

*"Our ability to sustain a positive outlook
means tending to it constantly."*
— Susanna Schrobsdorff

Now that we know the science of *why* optimism is important, let's talk about how to ensure we always have some around. By learning how to create optimism daily, we don't have to rely on the outside world to fill that closet for us. We can find optimism in ourselves.

Remember that choosing optimism is not enough. To remain an optimist, you must find ways to *create* it, but it's not as complicated as you might think. You simply have to become the change you're hoping for, and your body will take care of the rest for you. (More on this magic later.) Besides, as soon as you get going, the rewards will be so worth it that you won't want to stop. Becoming the change you wish to see does that to you—it makes you a better and happier person, and it's honestly quite addicting, but you'll soon see that for yourself.

Think of it as choosing to get in shape. At first, it takes work, and you have to deliberately decide you're going to get off the comfy couch and into your sneakers—you're basically forcing yourself out on that run. But soon enough you'll get in decent shape, and the results will make you want to go out for more, and before you know it, you'll feel so good that working out won't even be work anymore. It will be a part of life you can't live without.

Just like with six-pack abs and strong legs, optimism is earned. Your brain won't be fooled into thinking things will work out unless you begin to actively work in that direction.

We can *hope* to be optimistic, but to truly feel optimistic and show up with that energy, we need to activate that belief ourselves.

You can sit in the bleachers with your fingers crossed, hoping yelling will help improve the game, or you can get onto the field and start playing ball. One keeps you anxiety-ridden, the other throws you right into the game with real ability to make a change.

I am a climate optimist today because I've decided to be a living example of what that looks like. It doesn't mean that I hold every answer to reversing global warming. Even if I did, I probably wouldn't be able to do it all. However, I know some things, and with the little I know, I'm determined to do what I can. I'm choosing to be the change I wish to see in whatever capacity I can, and that attitude is what fuels my faith and keeps me going.

I'm not delusional. I'm very aware that right now we're not heading for a very bright future. I understand a majority of our habits and way of life are building toward anything *but* a sustainable and regenerative future. I know that, but I also know there are so many ways of doing things differently, and I'm determined to make those things my way of life.

I am not perfect in any way; no one is. We don't need a few people doing things perfectly; we need all of us doing the best we can, understanding that *imperfect* actually goes a long way. Being an *optimist in action* (OIA) is not about striving for perfection. It's about doing what we can, knowing we will probably fail from time to time, and still seeing those failures as new lessons that will keep us moving forward.

When I bring my own bag to the store, I'm well aware that the person behind me in line could very well grab ten plastic bags and wipe my positive footprint clean and then some. I understand that my efforts alone will not change the world, but they change *me*, and they are acts of making sure my optimism engine keeps running. I do these things (carry around a reusable cup, bring my reusable water bottle, say no to a lid at the salad shop, etc.) because they make me feel better and strengthen my belief in a different world. I don't just wish a new way of living was possible; I'm proving to myself that it is.

That is what creating optimism is all about. You understand you alone won't tip the scale, but you choose to show up anyway because you know your influence reaches so much farther than you could ever see. To be an optimist in action means walking the walk as much as you're talking the talk. It's about respecting yourself and your true desires to go the extra mile and make your dreams a reality—for yourself and for the world.

It also comes down to understanding that if you want to *see* change, you have to be willing to *be* change, and that, ultimately, it all comes back to you. The more you live with this mindset, the more you will see how you're planting seeds wherever you go, and the transformation that you'll begin to witness around you is going to be so reassuring and exciting that optimism is no longer a choice—it's inevitable. When you get to witness firsthand the changes you can inspire in your life and others, you will feel so awesome that optimism will become your middle name.

To be an optimist in action also means you understand things take time and this work we're doing is all part of a bigger journey. It comes with a certain understanding of the complexity of the picture we're trying to paint and how important it is we don't lose faith—that it's essential we keep our light shining and continue moving.

Being an OIA also means being *human*, with all that comes with fully embracing the complexities of who we are. That means it's okay to have bad days—terrible days too. It means we'll be heartbroken, angry, frustrated, and hopeless at times, but the difference that defines us—the optimist in action—is that we won't let those days turn into weeks or years. We will let these energies flow through us effortlessly to let them bring the messages of healing and reflection that we need, and then we will simply ask ourselves, "What would the optimist do?"

To live life this way takes practice, but the more we show up for it every day, the easier it'll get. What if all 8 billion of us started showing up as OIAs at the same time? What if every single person on this planet began asking questions, curious and excited to see what we would create if we just let go of the old and became more open to change?

Just thinking about it gives me goosebumps, and the optimist in me believes we will see this happening very soon, not just because I want it, but because I *know* how amazing it feels to live this way. And if you can promise me right now (a cosmic promise I hope you send to me) that you'll give it a try, then you're making me one happy gal.

Give yourself a chance at a life of excitement, transformation, and new possibilities. Please, my friend, choose the road of the extraordinary and become an advocate for positive change. The world will thank you, and so will I, and soon enough you will thank yourself—that's how great life as an OIA is.

The world needs people moving and asking questions. People who are willing to show up for the work and do what they can with whatever information they have now. We can drive from the West Coast to the East Coast in complete darkness as long as we have headlights showing us the next thirty yards of the road. With trust, our headlights will keep shedding light on the road ahead, and we can keep driving, for we have faith.

The road to a climate-just future will be anything but straight, and we must be ready for the bumps, traffic jams, and detours. We will have to act on our current roadmap and get ready to reroute from time to time. But if we keep our spirits high and find new fuel as the journey goes on, we *will* get there. Believe it and show up for the work—the journey—with passion and excitement, and I know we will get there.

Five Tips for Becoming an Optimist in Action

1. Slow down and start asking questions—to yourself, to others, and to your community. How can we unlearn and do better? Recognize that to change the world is to be willing to question what we know today.
2. Don't overwhelm yourself trying to do it all at once. Choose one thing you can change about your life today and then keep adding on. *Grow* with this journey.
3. Remember you're not alone. Positive actions in themselves are addictive, but positive actions *with others* become a movement. It's hard to leave a movement, so find your tribe and have fun.
4. Understand that humans belong in the cycle of life too. We're not just the problem; it's time to start recognizing how we are the solution. If we can move away from purely focusing on minimizing our negative footprint and instead look for ways to *maximize* our *positive* one, we can begin to see the role we really ought to play on Earth. Composting is a simple way of becoming that positive footprint—you're automatically reinserting yourself back into the cycle of life.

5. Remember you're only human. It's okay to have good days and bad days, so give yourself a break. Just make sure you don't make the breaks too long. Treat your optimism like any professional athlete would treat their sport, and you'll soon be in top shape.

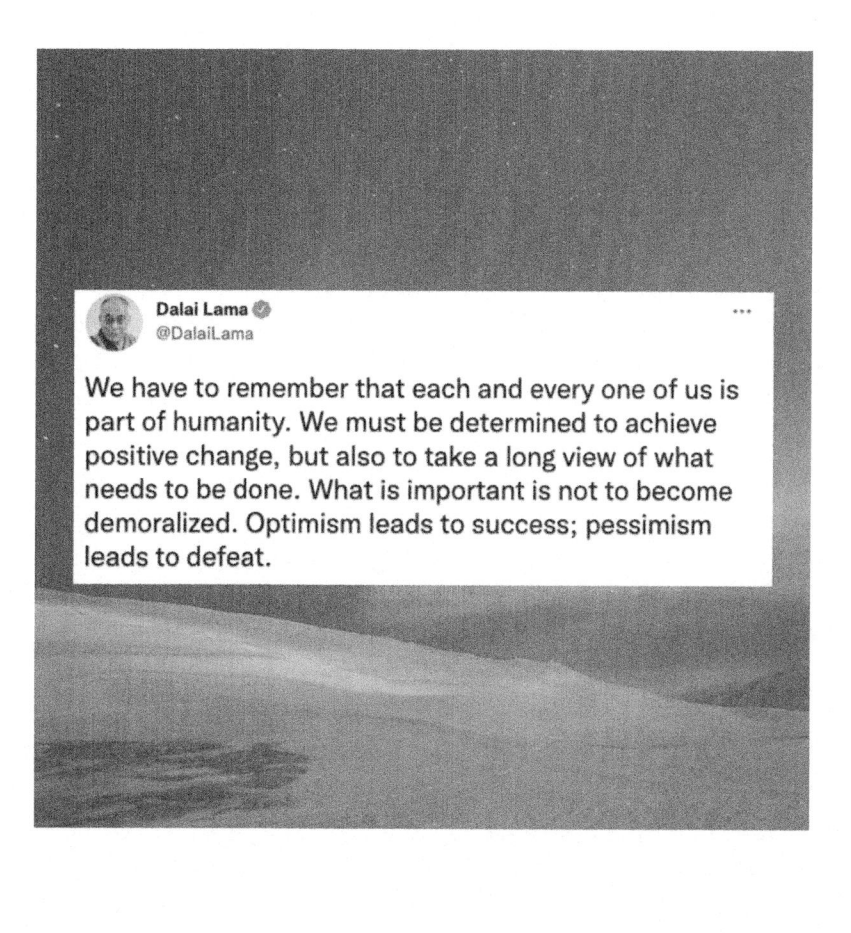

> **Dalai Lama** ✓
> @DalaiLama
>
> We have to remember that each and every one of us is part of humanity. We must be determined to achieve positive change, but also to take a long view of what needs to be done. What is important is not to become demoralized. Optimism leads to success; pessimism leads to defeat.

CHAPTER 53
Your Secret Sauce

In the last chapter, I hinted that to be an *optimist in action* is to live an absolutely awesome life. I didn't just say that so you'd be tricked into wanting to change the world—I mean it, and I have proof. What I've come to recognize, being an OIA myself, is that your body will actually reward you for the good work you're doing. It will say, "Hey, buddy, great job you're doing there; how about some feel-good hormones as a reward?"

That's right—being the change you wish to see in the world will help your body produce what are called happiness hormones, and I want you to get very familiar with them. I want you to think of them as your personal fuel and an important part of your journey.

Because we all know without fuel, there's no trip; if we don't have fuel in our system, we simply won't go very far.

If you drive a diesel car, that fuel is diesel. If you have an electrical car, then congrats—your fuel is electricity. But regardless of the vehicle, you need fuel. And if you're trying to be all smarty-pants on me, saying, "I'd prefer biking," well then, your fuel is whatever you put in your body—got it?

You need fuel to begin the journey, and you need to stop frequently to fill up. Fail to do so and your vehicle will run dry, and perhaps strand you in the most uncomfortable of places. Let's avoid that.

Given the vehicle is metaphorical and the trip we're talking about is your participation in cocreating a better world, what we're really talking about is *you*. This means you must understand to get any real work done (for yourself and for the world) you need fuel.

Many activists and change-makers tend to fall victim to what I call "angry activism." They believe that since anger triggered their commitment, anger is their only fuel. They get stuck. *As long as I remember how angry I am with the system, I can keep showing up for this daunting work!*

I know exactly what it feels like to be an angry activist because, as you know, this was me for many years. I used to believe that since I *knew* how unfair and broken the world was, I didn't deserve to feel happy until I'd fixed the problem. I learned to hate my own privilege and found it extremely difficult to participate in life's pleasures when I was well aware so many people had it so much worse.

Spending many of my young adult years angry with myself and the world, I know what this life is like, and I also know it's not working. At best you might lose a few friends and ruin some good parties, but worst case, you lose yourself, and the movement loses an ally forever. Being an angry activist, although powerful for kickstarting the fight, is not sustainable. We have to recognize the work is a journey, not a war, and therefore, we need something other than pistols and swords—we need fuel. The question is: What's the formula for balancing your anger?

Every person is unique, which means how we show up as activists (or should we say travelers) will look different too. And that's a good thing. If we were the same, we would have strange and not very productive movements.

What if all activists took to the streets and rioted but no one actually followed up and did something? "System change, not climate change!" Well, someone's gotta go change the system then, right? If we all turned into school-striking teenagers, we would have a lot of people missing class (and cool signs, for sure), but little actual change.

That said, we need everyone. We need strong voices, loving hearts, and analytical minds. We need people who are on the frontlines demanding justice—people who aren't afraid to speak up—and then we need those who are more reflective and inward-oriented to find the courage to start practicing this change in their communities and homes.

We need savvy people who study technology and science, writers and reporters who share what's been found, and "normal" people of all ages who are brave enough to question the truth and head into the unknown.

Activism takes many forms, and you get to figure out what role you want to play in all of this. You do play a role (or many); we all do. You may not have figured out what your role looks like yet. How can you make a difference in your life and community? How can you apply your unique strengths and contribute to a better world? If you don't know the answer right now, don't worry; you have plenty of time to figure it out.

Your role can also change. You can riot in the street one day and go home and write a letter to your local newspaper the next. Activism takes many shapes, and you get to choose what feels right for you, so if you've thought activism means quitting your job and joining Greenpeace, think again. So many opportunities to change the world are out there. You'll soon recognize just how many and how good it feels to contribute where you can.

But let's get back to your fuel. How do you find the fuel that works best for you? To put it simply, the fuel you want to create is composed of whatever makes you feel good. Not "good" as in instant gratification, but the kind of good that provides a stable and long-lasting foundation for a healthy and happy life. That is what a constant flow of happiness hormones does to you—they make you feel grounded, happy, and motivated, with an overall optimistic outlook on life. They reward you for getting the work done, but they also help you do even more, so the most beautiful part about them is that once you get them flowing, it is easy to create more. You need happiness hormones to find the motivation to continue on this journey toward a better world, but the work itself will also help you produce them. This is why being an OIA is addicting and why once you become one, you're not going to want to stop.

Get familiar with the four *happiness hormones* so you can better understand how to produce them and also *why* they're so important to this work. I hope by the end of this section, you will have a clear idea of the traveler you wish to be and how to create the secret sauce that will keep you going—happy, willing, and able—for decades to come.

Remember, without your fuel, you won't make it very far, so please treat this with the utmost respect. If someone asks, "So this woman, this crazy 'climate optimist,' is she asking you to *feel good and live a healthy life* as a way to fight climate change?" You can respond with a gratifying "Yes. Yes, she is."

We don't have a second to lose, so let's get going.

CHAPTER 54
Happiness Hormones

All right, let's get to know your awesome happiness hormones. You want to know how to seek out activities to keep them flowing (they're your fuel), but you will also see how becoming an active change-maker will produce all four.

Dopamine

Dopamine is commonly known as the body's "feel-good hormone." It's a type of neurotransmitter that our nerve system uses to send messages between cells. Many mental health disorders are linked to too little dopamine so it's very important we seek ways to produce it at all times. Dopamine is also the hormone that motivates us to do stuff, a productivity hormone that inspires work but also rewards getting work done. That's why I also like to call it the "Get Shit Done Hormone."

What does dopamine do? Dopamine helps you focus, keeps your mind clear, and motivates. It helps you learn, remember, and be productive, which is why you can get stuff done when your dopamine levels are high—you know, those mornings when you wake up rested and motivated to work? You also know the mornings when the opposite is true, when even though the alarm went off five times already, you don't want to do anything more than pull the blanket over your head and go back to sleep.

Dopamine is produced during a good night's sleep. That is why when you wake up feeling rested, you're more motivated to get to work, and it's more likely you will complete your tasks. In other words, never underestimate sleep. But you don't just create dopamine when you're sleeping. You can create dopamine by staying focused on your tasks. That, in itself, will help you sleep better at night.

In many ways, dopamine fuels itself. You need dopamine to find the motivation to tackle your tasks, but by crossing things off your list, you start producing *more* dopamine, which will motivate you to get even more things done. As you might see, being productive and making yourself proud can be quite a rewarding way of life. You'll sleep better, you'll be happier and more pleased with yourself, and you will get the important things done.

Naturally, when you *don't* do the things you're supposed to do—when you *procrastinate*—the dopamine levels are lower. The less dopamine you have, the less motivated you'll feel to tackle your to-do list, and the longer you'll keep putting it off…it gets harder and harder to muster the motivation to start.

If you've been stuck in the procrastination cycle, you know how frustrating it can be. You know you should just get to it and study for that test, clean the kitchen, or call to make a doctor's appointment, but for some reason, you just can't bring yourself to do it. The more you wait, the more frustrated you get with yourself, and the harder it is to actually take action.

Some of us are incredible procrastinators. I happen to live with one, and it amazes me how good someone can be at *not* doing something. I will, however, admit it has its benefits because whenever Arthur procrastinates, the vacuum cleaner comes out instead, and our apartment is almost always spotless. But being married to a master procrastinator, I've grown to understand it's one of those things where some people are prone to procrastination. If you're one of those, I'm sorry—you're just going to have to work a little harder.

However, we all do it from time to time—we put things off although we know we're fully capable of getting them done. How can we work around procrastination?

The absolute best thing to do in those moments is to kick-start your dopamine machine. Make a short list of things you want to get done and then start crossing them off one by one. Don't start with your big goal. Start small—that's how you build momentum. If "study for my final exam" is the goal with capital G, then identify some other tasks you can get going with right away, and then just simply do them.

Maybe the list could look something like this:

1. Do the dishes.
2. Put my pile of clothes away.
3. Call Grandma.
4. Pay rent.

Mel Robbins, lawyer, television host, author, and motivational speaker, has a powerful tool she teaches called the "5, 4, 3, 2, 1 Rule." It goes like this: you start counting backward from five to launch yourself off as your own rocket ship. When you reach one, you have no choice but to get right to what you were supposed to be doing. You can do it in stages too. Five-four-three-two-one—stand up. Five-four-three-two-one—walk to the kitchen. Five-four-three-two-one—do the dishes. Boom!

Have fun while you do it (may I suggest music or singing?) and observe as dopamine comes seeping back in. Use that to get to the next item on your list, and before you know it, you've created enough good juice to get to the big stuff. Practice this often, and you'll soon be a living dopamine machine.

Dopamine + Activism

When it comes to activism, you'll produce lots of dopamine. When you learn about something that doesn't sit well with you (for example, that a human eats about forty pounds of plastic in a lifetime) and you decide to act to see what you can do about it, your brain will release dopamine. Learning new things, trying new lifestyles, and growing as a person will make you feel really good because that's what dopamine does—it rewards you so you can keep going.

Serotonin

If dopamine is your productivity hormone, serotonin is your mindfulness hormone. Contrasted to dopamine, serotonin does not stimulate the brain; it calms it down and serves as a balancing force. Dopamine regulates motivation; serotonin regulates mood. You need both for a balanced and grounded life. In many ways, dopamine and serotonin work together. You want to add doses of both to your secret sauce. While dopamine is associated with rewards, motivation, and being productive, serotonin is associated with happiness, focus, and calm.[29]

How do you get more mindfulness hormones flowing? You may not be surprised to learn that activities like yoga and meditation will do the trick, but there are other ways too. If you struggle with sitting still for too long (I'm one of those), seek similar activities that might be more your cup of tea. How about a quiet walk in nature or simply taking a few quiet minutes to drink your coffee on your steps in the morning sun?

Serotonin levels are also linked to exposure to sunlight, which is one reason (if you live in those parts of the world) you could experience seasonal depression (seasonal affective disorder). Your mood dips when the days are shorter because you simply don't get enough sun. This is why finding ways to step out during the day, even if for just a short walk, can be tremendously helpful to your mental health. Don't waste all those bright hours behind a screen or inside your home.

Low levels of serotonin can also lead to poor sleep and digestion, which will make you feel more icky and stressed. As they say, "Happy gut, happy life." I know that to be the truth. Having suffered from a troublesome gut for many years, I know what it's like to live with a grumpy tummy. It's not fun. Learning to eat better and regulating stress has been essential for my wellbeing. I can't tell you how different a day is when my gut is happy compared to a day when it's not.

Serotonin + Activism

As activists, I know it's easy to feel drawn to action-driving hormones like dopamine, but don't be fooled—we truly need both. Although activism has a lot to do with getting things done, it's also very much about slowing down. You need to create space to reflect and go inwards. That is how creativity sparks and where you come up with new solutions and ideas. You also need it to find the courage and optimism to keep going, because with more serotonin, you will feel more at peace with yourself and the world, and that is how you grow hope. Remember you don't want to be a chicken panicking and running around. You need to be grounded and optimistic to continue to show up for the work. Plus, the more grounded you are, the better you sleep and the more committed you'll be to your actions. This is

---

[29] simplypsychology.org

why you need serotonin and dopamine to work as a team.

As the world continues to spin faster, seek out activities that help you reduce stress. Ground yourself through meditation or walks in nature, or seek other pleasures that help you slow down. Do these things and you will sleep better, which will let you wake up feeling more rested and ready to kick off the day. That feeling will spur your commitment to making the world a better place, and those actions in themselves will make you feel more fulfilled and grounded. And do you know what that will help you do? Sleep better at night. See what I'm getting at here?

There are many ways to produce both dopamine and serotonin. However, I highly recommend you manage your physical and mental wellbeing with great respect. Sleep right, eat well, and treat your body as the temple it is. The world needs you to be as happy, healthy, and grounded as you can possibly be.

Oxytocin

This one is for those who tend to think they're superhuman and can do it all on their own. Oxytocin is all about the *community* piece of activism, and soon enough, you'll understand why teamwork truly does make the dream work. It's not just about mixing talents and ideas. We need community for activist fuel too.

Oxytocin serves as the bonding hormone in your body. I'm yet to be a mother, so I don't know what it's like to breastfeed a little one, but they say when you do, your oxytocin levels run high.

Our body knows we need other people to survive, so it will reward us for doing things that bring us closer. That is how we *bond*.

A parent needs to feel the bond with their baby to nurture and protect it, but this hormone is important beyond raising offspring. We need to keep producing it throughout our lives. As humans in a big, confusing world, we need one another to build community, overcome difficulties, share ideas, and create new things.

That is why we need other people and why too much time in isolation can make anyone go mad.

Oxytocin is commonly known as the "love hormone," and considering how it's produced, you can see why. Having high levels of this hormone makes you feel loved, and it creates a connection with other people. You begin to feel like you're part of something bigger, be it a family or a group, and that, in itself, sparks optimism and hope. Higher levels of oxytocin increase trust, which makes you feel more at ease, allowing you to let down your guard. As you know from the science of optimism, you need to lower your guard to widen your horizon and find the courage to dream. That is why the feeling of trust and community is essential to any serious activist.

Oxytocin + Activism

On your make-the-world-a-better-place journey, I want you to think of oxytocin as the people who ride with you in the car. If you drive alone, you're going to have to take a lot of breaks because no one can keep driving forever. But if you have others with you who can take turns, you can jump over to the passenger seat from time to time to take a well-deserved nap. You can snooze away and dream of better days ahead, happily knowing you're still on your way.

Activism is a lot like that. There will be days when you wake up and have zero interest in changing the world. Some days you will just want to binge-watch Netflix and tackle a pint of (dairy-free) Ben & Jerry's in bed. It's totally okay to have those days. In fact, you need them too. If you never let yourself rest and recharge, you won't get very far. But if you're not alone on this trip of yours, if you have a community of people you *know* are as passionate about making the world a better place as you are, you can rest with a good conscience knowing someone else is doing the legwork that day—that even if you're taking a break, the movement is still moving forward.

The best way to add oxytocin to your fuel is by joining a community, either a local grassroots group or one that speaks to you online. Connect with people who care about the same things you do and join them in their work. Learn from other people and also share what you're learning. Make it your mission to establish a network so strong there's no chance you'll fall through the cracks on days when things get hard—they simply won't let you.

We are social creatures. We belong in a community because that is how we thrive. Paying attention to our oxytocin levels will tell us just that. We feel good when we're part of something bigger because our body *knows* how important it is. Whatever you do, don't go it alone. Find your fellow travelers and have fun together on the journey.

Before you panic and start to obsess over your lack of social interaction, find solace in knowing you can actually create oxytocin without other people. Start with a small dose you can create at home, all by yourself, and then use it as fuel to go seek more.

You don't need to fall in love or have a baby to produce oxytocin. You don't need to show up in a room full of people to find community and support. To begin, you can create love and support for yourself. Acts like singing, dancing, laughing, and small pleasures like taking a hot shower or treating yourself to a good piece of chocolate will produce oxytocin too. You also don't need physical interaction. Simply calling a friend and chatting as you both make dinner can easily do the trick, and these simple acts of bonding will warm even the most quarantined heart.

If you *do* have a partner or parent around, don't hold back on the intimate pleasures life is so full of. Kissing, hugging, sex—even eye-gazing (a personal favorite)—are amazing ways to produce oxytocin, and remember you need it for fuel, so make sure to get a lot of it! The best activists out there will be those who take on life with all its glory and use it as the fuel to get the important stuff done. At least that's my opinion, but feel free to form your own.

Before we leave oxytocin, I want to share the most powerful source of this hormone—small acts of kindness. Some studies show the highest feeling of happiness comes from selflessly helping other people.[30] I don't mean doing someone a favor because you expect to get one back, but giving kindness simply out of love, like helping an elderly person cross the street or carrying their groceries to their doorstep. Doing these things *will* make you feel amazing, and you have every right to feel that way. It's not your smug self; it's the oxytocin.

We're meant to help each other. We're meant to live in community and be of service to each other, no strings attached. Our reward system helps us

---

30 Cohut, Maria. *Medical News Today*. July 16, 2017. https://www.medicalnewstoday.com/articles/318406. Study published in *The Journal of Positive Psychology*: "Happiness comes from trying to make others feel good, rather than oneself." March 8, 2021.

see this, and I think it's about time we follow the feel-good hormones and use them as clues to how to live fully—and not just how to live, but how to live out our unabridged service to our community and the world.

Endorphins

Last but not least, you have the hormone you might be the most familiar with—endorphins. Often known as the body's pain-killer or "happiness hormone," this hormone is a nice treat you produce when your body reaches a certain pain threshold, such as when working out. It's commonly known as the "runner's high" because, after a certain distance, endorphins get released so you can keep running. The reason you feel "high" is because it acts on the same part of your brain as opioids, such as morphine, hence it makes you happy.

Endorphins are produced to strengthen your immune system but also to increase pleasure. As Donna Hartley writes, "Since humans naturally seek to feel pleasure and avoid pain, we're more likely to do an activity if it makes us feel good. From an evolutionary standpoint, this helps ensure survival."[31]

In other words, endorphins help us do things that are good for us we might otherwise avoid, like chasing animals for miles to get food like our ancestors did.

Endorphins + Activism

Lots of activism has to do with moving your body—showing up for a beach cleanup, taking the bike instead of the car, going hiking to connect with nature, or carrying around your slightly heavier water bottle to refill so you don't have to buy a plastic one at the store. Yes, being an earth hero takes some physical activity, but since you're not made to sit still all the time, this kind of lifestyle will only reward you. You will feel healthier, happier, and more alive, and soon enough, it won't feel like work at all—it'll simply be the lifestyle you can't get rid of.

As mentioned in Part Two: Choosing Change, we tend to fuss a lot when we have to get used to new habits, so you might make a big deal out of having to bike to work, or make excuses instead of showing up for

---

[31] Hartley, Donna. "Endorphins." August 10, 2020. https://www.strandfitness.com.au/endorphins/

that clean-up. However, hijack your dopamine machine and just do it, and watch your mind and body shift as you slowly get used to new things.

You don't have to be a runner to infuse your body with endorphins. You don't have to reach a runner's high to reap its benefits. Activities like dancing, laughing, sex, and simply taking a break to watch a drama on TV have also been shown to produce this happiness pill.[32] Benefits of endorphins include reduced anxiety, improved self-esteem, enhanced immune system, and increased sex drive—rewards that, again, incentivize you to do more of the activities that produced them in the first place.

That's the wonderful thing about happiness hormones. As you begin to produce them, you will automatically seek out exercises that produce more. It's like a wheel of happiness and health spinning along and fueling itself. All you have to do is commit to that lifestyle and get started. That is the secret ingredient to your secret sauce—a life lived in happiness, health, passion, excitement, and lots of joy.

Stop procrastinating and start getting things done. Set goals for yourself and commit to a life of continuous growth and learning. Slow down and invite mindfulness into your life, and find ways to be more here for each moment. See your neighbors, respect Mother Nature, treasure the now, and tell yourself you are allowed to feel good about being alive. Be here, seek joy, and keep finding ways to do good—the happiness machine will be there to reward you.

If you've been denying yourself simple pleasures, give yourself a break. You're trying to change the world so you better give yourself permission to be rewarded for your efforts. Your body will begin producing all these hormones to thank you, so don't push that back—embrace it with all you've got. It might seem strange to be so aware of climate change and at the same time start feeling really good about yourself and the world, but that is exactly the fuel you need to keep going.

Why do you think people stay in philanthropy work for so long? Because it makes them feel good, that's why. Now you have it, the secret to heroic work—start helping others and you are going to get rewarded and rightfully so.

Chances are once you start implementing these small changes in your

---

32 verywellmind.com

life, based on all the new things you learn about yourself and the environment, you'll start to look at life through a new lens. You will begin to feel like you matter and maybe there's hope after all. Remember this feeling—the pride, happiness, motivation, and excitement—is the fuel you need to keep your engine running, so receive it with joy and keep producing more.

You're not separate from nature; you *are* nature. You're not separate from the system; you *are* the system. You are the change you're looking for, and the sooner you can embody the *feeling* you want the world to be filled with, the sooner you can manifest that feeling in the world. Remember, taking care of yourself *is* taking care of the world.

Happiness Hormones Cheat Sheet

Dopamine

- Feel-good hormone
- Learning, memory, motivation, and productivity
- Getting things done and learning new things
- Enhances sleep cycle

How Dopamine Shows Up in Activism:
- Rebel sanity—doing what you can to make a difference ("being the change")
- Learning new things and sharing—constant growth
- Implementing small changes—builds momentum and fuels optimism

Serotonin

- Mindfulness hormone
- Yoga, meditation, walks in nature—slowing down
- Peace of mind, grounded, optimism
- Helps you see things more clearly and practice love and gratitude

How Serotonin Shows Up in Activism:
- Slowing down provides time for reflection and *retruthing*
- Deeper connection with people, self, and nature
- Feelings of optimism, peace, and hope equal fuel for action

Oxytocin
- Love hormone
- Bonding, trust, belonging
- Enhanced through eye-gazing, sex, dancing, massage, hot showers, and small acts of kindness
- Makes you feel like you're part of something bigger

How Oxytocin Shows Up in Activism:
- Community building, a feeling like we're not alone
- Bonding over joint passions and interests
- Collective action makes us feel better

Endorphins
- Pain killer hormone
- Euphoria/runner's high
- Strengthens immune system
- Increases joy and happiness

How Endorphins Show Up in Activism:
- Gardening
- Carrying (heavier) zero-waste products
- Biking and walking instead of taking the car
- Beach cleanups
- Grassroots activities and organizing

CHAPTER 55
Optimism in Practice

I hope you understand by now that optimism isn't just a choice; it's a lifestyle. And more—it's not something you get to choose or not choose; you must create it for yourself. Like working out, you won't stay in shape unless you commit to ongoing practices that will keep you motivated, hopeful, and excited about the challenges at hand. It's very much a "glass half empty or half full" kind of attitude, but there's more to it than the choice alone. Once you activate joy, once you make the determined decision that there is hope for the future and you will make it your daily mission to make sure there is, you've begun creating optimism on demand.

Like any good wheel, this one fuels itself. Choose to be the change you wish to see in the world and feel happiness hormones spread through your body. Pay attention to how they make you feel and how they motivate you not just to keep going, but to do even more.

What do you see now that you didn't see before? What do you believe to be possible now that you previously barely dreamed of? Trust those whispers of inspiration because they may very well be telling the truth. It's not just about feeling good; it's about raising our odds in everything we do.

You are a powerful vessel for change, and that is what I want you to remember when you wake up every morning. Sure, your actions alone won't change the world, but you will be one more soul adding your powerful energy to the collective, and the more people who keep raising those vibrations, the more we'll be able to get done.

Here's a visual I created to make it crystal clear how this wheel of optimism works. This, my friend, is what it means to be an *optimist in action*.

> HOW TO CREATE OPTIMISM

Activism

Optimism

Happiness

#OPTIMISTINACTION
THE CLIMATE OPTIMIST HANDBOOK

Commit to becoming an activist however that looks to you. Remember you don't have to create a witty sign and show up for a climate march to be one—activists take many shapes and forms. Maybe your activism is more subtle and bleeds into everything you do. It can be quiet acts of slowing down, listening, asking questions, and starting conversations. Or your activism could mean you decide not to buy any new clothes for a full year—that's a lot of power right there.

Once you begin to practice flexing your activist muscles, you will recognize everything you do is activism as long as there's some dose of intention involved. Choose to be an activist because it will be rewarding. Happiness hormones will begin pumping, and with the secret formula to your unique fuel flowing through your body, you'll feel so great there's no way you won't want more. Not only will you feel great, but you will begin to see yourself and the world differently too, and with a fresh pair of eyes, who knows what you will explore?

This is what optimism on demand looks like. This is what it means to be an OIA. You recognize the only hope is the one you choose to create for yourself. These acts will make you feel better, and hence, more optimistic—optimism you use to take action, and in doing so, produce even more actions.

More, more, more! The wheel of "Yes, and…" can keep on spinning for a lifetime if you let it—it's the ultimate renewable energy machine.

The only question I have right now is…how will you activate it today? What is *one* thing you can do to be the change you wish to see and set the wheel in motion? Don't overthink it. Start small. Simply begin and experience the magic.

CHAPTER 56
Why Climate Scientists Are Optimistic

Believe it or not, some strange people out there decided to dedicate their lives to studying our planet's fall, tracking every change in our climate-changing world, and signing up for a life of worry and despair. We owe these people so much because, without them, we wouldn't know to act in the first place, and we would have very little idea of what to do now that these scary moments are actually here.

The people I'm talking about are the climate scientists of the world. I believe it's about time we see them as the heroes they truly are. Despite decades of attempts to get the public (and politicians) to listen, they keep being ignored as their well-researched reports get either buried or challenged with one ridiculous argument after another. If you've seen the movie *Don't Look Up*, you get the picture. Imagine trying to warn humanity of crises so big none of us can even fathom them, only to continue being laughed at or shut down. How climate scientists even kept going, I don't know, but we should all be thankful they did because now we have a pretty clear idea of what is going on—and what we need to do to fix it.

What's most fascinating to me is that despite how *aware* these people are of the crisis at hand, they tend to be a fairly optimistic bunch. They keep putting out new proof and data because they want us to *act*, and they wouldn't try so hard to make us act if they didn't think we actually had a chance. When you think about it, climate scientists might be the most optimistic bunch of people we have!

Now, I also know that many climate scientists are finding it harder to stay motivated and are losing themselves to hopelessness and despair, and who can blame them? If decades of work still leads to nothing while you watch the world do exactly what you predicted it would do, only *faster*, who wouldn't lose hope? I know this is true because I had a climate scientist from South Africa join one of my online workshops once, and she said afterward that her people—the scientists—may be the ones who needed this message the most. It honored me deeply that a scientist found meaning in my climate optimism, but I also understood why. Sometimes you simply need someone else to give you the simple tools (encouragement) to keep going.

However, climate scientists are not just optimistic because they think we don't have a choice. They remain so because they *know* we have a chance to fix this. In 2017, Paul Hawken, a world-renowned environmentalist and celebrated author came out with the book *Drawdown*, consisting of a list of the hundred most efficient *solutions* for reversing global warming. To write his book, he talked to scientists from around the world. He was careful to exclude anything that wasn't already proven with results that could be measured and compared. He wanted to be able to rank these solutions and give us—the people—a true guide to the solutions we're already aware of and we can start working on right now. In other words, the book is not filled with wishful thinking, but with meaningful actions that can be implemented today.

Scientists are not optimistic because they hope for things to work out. They continue to show up for their work because their science shows it *can* be done. However, we need to act now, and that is what they keep telling us. *We know what's going on, and we also know how we can keep the worst from unfolding, so let's get serious about the future we want and save ourselves from our demise.*

The message is pretty clear: We need to look at the science, *trust it* (because we have the proof), and decide we care enough about our life here on Earth to act. It's not just about people in power, nor is it about what's happening in other countries. This issue hits home for all of us, and we need to see that and bring the power back where it belongs—in our own homes and hearts. We have a chance to fix this because we all have a chance to act, and that's what it comes down to—we all need to act. Quoting one of my favorite climate scientists again, Katharine Hayhoe: "As optimists based in science, we have to recognize that the future truly is up to us."

CHAPTER 57
Diamonds on the Soles of Her Shoes

I have a few songs that serve as my personal anthem. Paul Simon's "Diamonds on the Soles of Her Shoes" is one of them. For one, it's so fun and uplifting it can butter up any mood. Just thinking about the intro sung in Zulu brings me back to walks in Central Park when I had the song blasting in my ear and felt like nothing could stop me. With the sun on my face and Manhattan's skyline peeking through the lush trees, I felt so big and special it was like floating on a cloud of magic. How can life *not* be magic when you walk around with diamonds on the soles of your shoes?

I'm sure you have one of those songs too, one that's hard to listen to in public because all you want to do is dance, and you're worried you'll look like a complete fool if you do. On a few rare occasions, you will spot someone on the subway who is so deep into their music that they couldn't care less about the world. Eyes closed, body beating to the (for the rest of us) nonexistent music, lip syncing to the lyrics of the song. I love those people. They remind me that truly being here is all that matters. Because who gives a shit about what other people think when you're that attuned?

What I've sincerely concluded is life is too magical *not* to walk on diamonds, at least figurative ones. (And if you have money enough to walk on real diamonds, please make sure they're ethically and sustainably sourced.) When the song comes on, it's like a nudge from the Universe that I'm here to play, I'm bigger and more magical than I think, and if I walk around on diamonds, I will notice all the extraordinary things in life. It's a call to level up my step and remember my importance in this world because, believe it or not, every step we take matters.

Having diamonds on your soles is a metaphor for me. The actual song is about a boy who falls in love with a rich girl, and I have yet to call Paul Simon to ask him what he was trying to say. But it doesn't matter.

The song has already changed my life many, many times. Walking on diamonds is not even just a metaphor anymore—it's a mindset, a way of life. It's recognizing you sparkle so much that the world around you has to sparkle just as much, and that's why you strap on those shoes. You recognize when you walk around with sparkles on your feet, you'll spread fairy dust around and make the world a more sparkly place for all.

Gabby Bernstein, author, motivational speaker, and podcast host, has a mantra that goes, "I bring light with me wherever I go." To me, that light is carried in my heart but also on my feet. My feet step into a new room. Whatever my feet touch—the floor, the ground, the grass—is also connected to those around me. We stand on the same foundation. We're rooted in the same earth. Like trees that build networks of mycelium underground, a web of life so magical we can barely fathom it, so are we creating a network of energy wherever we go.

In traditional Hindu society, going barefoot is of prime importance. It signifies respect for the earth upon which people walk, the sacred mother of all. I'm not suggesting you go barefoot (especially if you live in colder places than India), but you can consciously reflect on what footprint you're leaving behind, not just your ecological one, but your energetic and spiritual one too. What energy do you bring with you when you enter a room? What message are you sending to our spiritual mycelia as you walk around this world? Are you walking around with love, gratitude, and self-respect? Do you love and see yourself enough to love and see other people? Do you rejoice in the here and now and recognize these precious moments, the *present*, as the gifts they truly are?

And if not, can you maybe try putting on some different shoes?

When I have my diamond shoes on, I see magic. I see heart shapes everywhere—in rocks on the dirt path, in dried gum on the street, and in strange formations on the wall. When I see those hearts, I know my magic shoes are working. I know they are because I'm paying attention.

I also recognize the truly special moments, like a child squeezing her parent's hand a little extra when walking through a crowd—a sign of trust and comfort. Or two elderly hands touching fingertips across the table at a restaurant—a sign of love.

Our world is filled with magic because we *are* magic. Like magic, we continue to create magic—all the freaking time. However, sometimes we have to put on our sparkly shoes to step into a scene with the right attitude to see it. When I walk around with diamonds on the soles of my shoes, it's not because I want to be better than anyone else or sparkle more; it's because I want to see magic, and I can only do so by recognizing that I am magic too.

You are magic, so get some diamonds on your shoes and start paying attention.

CHAPTER 58
Magdalena from the Future

Everything is different now, yet undeniably the same. Magdalena doesn't think about it that often because it's incredible how fast you get used to something new. However, sometimes the memories creep in, memories of a different world.

It wasn't that life was necessarily bad back then, but she often marvels at how resistant everyone was to change, especially when *this* was waiting around the corner. If only we had known, would things have happened faster? And with less pain? Maybe not as many people would've had to suffer in the transition. Maybe not as many species would've had to go extinct.

It's Thursday and Magdalena is headed to town to meet her friend for coffee, something they've been doing every Thursday since their spouses passed away. It's hard to be alone, but not unusual at their age, and besides, she counts her blessings every day that she gets to grow old in *this* world.

Magdalena closes the door behind her and places her hand over the docket on the wall. A message on the screen tells her the house has entered power-saving mode and will begin to direct energy to the backup system. It's sunny out so the computer says the battery will be loading 75 percent from the sun and 25 percent from food waste. Great, she thinks, because she hasn't been creating that much food waste in the past few days.

The brisk morning air feels heavenly in her lungs, and Magdalena takes a few deep breaths, eyes closed toward the sun, before starting her walk to the ThruLine. She knows her friend is waiting, but there's no rush. In this weather, she'll happily wait a few more minutes. Besides, the train comes so often these days there's really no risk of missing it, so why rush?

As she walks to the station just around the corner, she allows herself to travel down memory lane. Remembering the old days feels important so they can continue to choose consciously moving forward. They must remember where they came from so they can find pride in what they've created, and ensure they pass on this legacy of consciousness to future generations.

It's a new era, and things have changed. Not just infrastructure, society, and businesses have changed, but something deeper as well. The collective consciousness feels different these days, and you can see it in people's eyes if you pay attention. People have a spark that definitely didn't exist before, but is hard to describe if someone were to ask.

Magdalena laughs as she realizes why it was probably so hard to understand back then. Even if someone would've known, how do you articulate that message to the masses?

"Oh, everything will be different because there will be this spark, you see, in people's eyes. And when you see that spark, you'll know we've made it."

Ha. People would've thought you were a loony. But if you talk about it today, people understand—not everyone, of course. In the history books, you'll read about the "Renewable Era," the bold collective moves of governments around the world, and the people in power who finally decided to come together to make this right. You'll read about the monumental courage of the people who had to let jobs, structure, and safety go to build something better. You'll read stories about heroes from all walks of life who, instead of seeking fame and entitlement in the old world, chose to mobilize and lead people into the new. And there will be example after example of courageous people who were the backbone of the revolution but never made the evening news.

The thoughts bring forth a smile. It's been a hell of a ride, hasn't it? There's so much courage to be proud of, so much hardship and curiosity to be remembered and celebrated.

Magdalena's thoughts get interrupted by the train, which seems to have appeared out of thin air. She's having a hard time getting used to it—noiseless vehicles seeming to appear out of thin air. At least back in her day, they made enough noise to give you a heads up. Well, she wouldn't trade that heads-up for any pollution in the world!

The doors shut silently behind Magdalena, and she sits down in one of the empty seats. A young person across the aisle is reading a book and the title catches her attention. *We Will Forever Be Grateful for Those Who Had the Courage to Question Everything.*

Maybe it's something they make them read in school these days? Magdalena makes a mental note of the title and decides to try to find it later. It's important to know what they teach the kids to ensure they don't start going backward, only forward. We're a young world with lots to learn, but the opportunities are endless—that she is absolutely sure of. As long as they remember the truth behind the revolution, as long as they know it all came down to the courage in people's hearts and the unified strength of community across borders, they can keep moving forward into even better, fairer, and greener worlds.

The ride is short on the fast-moving train, and as the doors open at the downtown stop, the sound of life instantly fills the car. Porcelain clinking, people talking and laughing, a street musician playing farther down the mall, and mixed in with it all—birdsong!

Magdalena loves a lot of things about this new world, but what she loves most is the city. The city came back to life in ways no one thought a city could, with culture, life, greenery, and above all, a sense of community.

She spots her friend by the gates down the platform, eyes closed in the sun as hers had been only minutes ago. She walks over to her with new life in her step—it's time for the weekly gossip and catch-up session.

• • •

This is my story about Magdalena. I don't know how far into the future it is or how old Magdalena is exactly—it's hard to tell because when describing her she feels younger than anyone I know today.

I also don't know if this is what the world will look like. I'm not an engineer or scientist, nor am I a fortune teller—I don't hold any insights into the future that other people don't, but I love to imagine. I love to tap into my childlike heart and ask, "What world is possible if we only keep some room in our hearts for the unimaginable? What future could we find ourselves in if we only dared to believe?"

I know change isn't easy, but it's inevitable. In fact, it's the only thing we can really count on. You might have an idea of the future we're currently headed toward, a future with melting icecaps, disappearing rainforests, oceans filled with more plastic than fish, depleted soils incapable of growing food, deteriorating health, increased cases of asthma, wildfires—the list (unfortunately) goes on.

And that's not even the *future*. That's what we're looking at right now. What will this mean for life on earth? We can't know with 100 percent certainty, but scientists have their suspicions, and it's not going to be pretty.

This is the change being handed to us on an uncomfortably heavy platter, and unless we find something better to counter with, that's what we're getting. So, let's get serious in our search for what we *do* want, and let's get going on *that* journey. Again, let's get serious about choosing change.

Do you want to be a hero? Do you want to be part of writing history? Do you want to have an incredible story of will and bravery to tell your children and their future kids? Then buckle up because we're in for one heck of a ride. It's time we get excited about these challenges we're facing and start to see them for what they really are—ways to come together as a world and decide we can do better. The journey we're about to embark on is one of love and regeneration, where the biggest change isn't what happens around us, but in our hearts. Because if we want to change the world, we have to be willing to change *ourselves*. We are the world, so the world will go where we go, and that's one powerful thought once you let it sink in.

A climate-just future is our destination, but what exactly that will look like is up to us to figure out. And here's the caveat—we can't know unless we start approaching that destination, and we can't do so unless we decide to finally take off. You don't have to know how we will solve the climate crisis and build this new world, but if we get going, we can learn. Just like Google Maps can't reroute you to a better way before you actually start driving, we can't get closer to our goal until we start making our way there.

Will it be scary to let go of the old and rebuild society as we know it? Sure, it will, but I believe we have what it takes to do it. I believe we have it in us to make history and cocreate a better world. I just hope you start believing it too.

"Perpetual optimism is a force multiplier."
— Colin Powell

"Optimism is the faith that leads to achievement.
Nothing can be done without hope and confidence."
— Helen Keller

"I believe any success in life is made by going into
an area with a blind, furious optimism."
— Sylvester Stallone

"For despair, optimism is the only practical solution.
Hope is practical. Because eliminate that and it's pretty scary.
Hope at least gives you the option of living."
— Harry Nilsson

"Optimism with some experience behind it is
much more energizing than plain old experience
with a certain degree of cynicism."
— Twyla Tharp

"Hope is often misunderstood. People tend to think that it is
simply passive, wishful thinking: I hope something will happen
but I'm not going to do anything about it. This is indeed the
opposite of real hope, which requires action and engagement."
— Jane Goodall

PART FIVE
CHOOSING EMPOWERMENT

"The thing about climate change is that you can either be overwhelmed by the complexity of the problem or fall in love with the creativity of the solutions."
— Mary Heglar

CHAPTER 59
Activate

At this point in the book, you should have a good understanding of the challenges we're facing and how important *your* part can be in co-creating a better world. You understand to create long-lasting change, you need optimism to fuel you, but you also know optimism is earned. By finding the courage to question what is and aim for even better ways of living on Earth, you get to be the hero we need. We can't wait for one brilliant mind to save us all because it simply won't be enough. We need thousands of new ideas and solutions backed by people who are willing to trust we can create something better. If smarter products come to market but no one buys them, what's the point? If new innovations are created but Congress doesn't have the power to use them, how can they help?

The truth is there are already so many cool products and services made by people who genuinely care about making this world a better place. Those people need our support to grow and show the marketplace we can do things differently. If no one buys their products, they will continue to be killed by the giants that can produce large quantities at much lower prices. All the new ideas need are some curious consumers who choose to believe in something better and who *care* enough to give their products a chance. Be that curious customer—have the courage to fuel a positive economy and pave the way for others to follow.

You may not know which products I'm talking about at first, but put your awareness eye on and soon enough you'll see them online or on the shelf when you're shopping. You'll see something a little different, a product promising healthier ingredients and eco-smart packaging. Do something crazy and give it a try.

Brilliant ideas for creating energy, storing energy, and sharing energy between homes are being presented. As are ideas on how to grow food more efficiently to save land and water, how to use dead spaces and rooftops for urban gardens, and how to grow kelp forests in the oceans to restore its ecosystem while simultaneously capturing carbon from above. Smart solutions for how to create an eco-efficient and healthier world, as well as projects for how to *regenerate* the Earth and *restore* the damage that's been done are being brought to light. Many of them have already been set in motion and are making a positive difference as we speak. Support these efforts. Be curious enough to want to learn about them and share what you learn with family and friends. Spread the good news and start telling the story that we actually *can* do this—just look at everything that's already been done. When you focus on the positive change already in motion, you will see that not only is a better world possible, but it's our only choice.

I want you to recognize you have a part to play in all this. Even if you're not an entrepreneur with a cool idea or an engineer with the opportunity to create new solutions, you can do so much to help. By getting involved in local politics you can push for new initiatives and show your elected representatives what you care about. By talking about these solutions (and why they matter) with relatives and colleagues, you can get more people onboard in believing in a better world. That is where your true power lies—*you* have the power to *believe* and give these new ideas and initiatives a chance. So be a supporter, be a believer, dare to be a little different, and push for an even better world.

Often you might feel like you have to sacrifice in order to be this champion, but I ask you to challenge that thinking. Driving less to cut down on your CO_2 footprint might seem like a loss at first, but think about the long-term goal. If continuing to pollute the Earth will mean there won't be much of our world left to enjoy in the future, is it really that much of a sacrifice? Besides, the instant benefits might start to appear soon too. If you don't drive so often, but instead plan your trips to the store and become smarter with your stops and visits, think about all the time you will gain. Suddenly, you'll have more time to read a book, cook a stew, or simply sit outside and listen to the birds while having breakfast.

The gratification that comes from a life of doing *less* might surprise you, but as it does, use it as a reminder to change your narrative. You're not missing out at all, are you? You're only pivoting to find even better ways to live.

You *are* the change, and you get to activate that change at any time. My hope is you don't wait too long because the world needs you, and the life you'll start living will be so much better. I don't have to sell you on all this because you already know the benefits of being an optimist in action.

In the coming chapters, you will learn how to build empowerment from within and gain some tips for how to be the smartest, most grounded change-maker you can be. Let's get ready for some positive change.

> *"The greatest threat to our planet is the belief that someone else will save it."*
> *— Robert Swan, OBE, First Man to Walk to Both the North and South Poles*

CHAPTER 60
Planting Seeds

I paced up and down the street trying to figure out what to do with myself. I was angry and upset at the same time, but I also understood I was acting out like a child. My mom, aunt, and a friend had come to visit me in New York for a week, and we'd had so much fun as I took them to all my favorite places. However, right now, they were seated at a burger restaurant in Hell's Kitchen, and I was a couple of blocks away, furiously pacing. I had simply stood up and stormed out without a word, leaving myself in this childish position.

One reason I'd picked the restaurant was because they had some incredible plant-based burgers. The bean and mushroom patty was my absolute favorite, but there were many, and I was so excited to introduce them to the crew. Naturally, none of my guests opted for the vegan options. They all chose different kinds of meat instead. Being so early in my vegan journey, I had to comment, and so the conversation was sparked. Should we really be eating meat?

My aunt brought up some solid points about cows grazing and how it was good for the land, and I argued most animals weren't raised that way, especially in the US. With two fair points made, I could've left it there, or I could've at least continued with curiosity and understanding, but it was way too early in my journey for that. Instead, I instantly had an emotional outburst. I could feel the anger rising in my chest and the tears start to burn at the corners of my eyes. Not wanting to make a scene, I simply stood up and left, not even looking back at their (I assume) surprised faces.

I contemplated leaving altogether. They had their own key to the Airbnb, so I could just meet them there later. Despite them being my family, I was very angry. I felt alone, like no one in the world understood me, or that no one understood *them*—the animals I desperately tried to speak up for. How would I ever be able to change the world if even my own family didn't listen and understand?

I had been vegan for a little over a year, and so far, it had been quite an exhausting journey. People would confront me at parties, and I always felt the need to defend myself and educate, like my whole life suddenly turned into a faith-like experience, and everyone was questioning my faith. I didn't mind sharing my experience, but I never felt like I actually made a difference. It never felt like my words moved or sparked the desire to change in anyone. They always nodded a little and then changed the subject, leaving me feeling exasperated and powerless. I knew what I said hadn't hit home.

Eventually, I decided to head back. As I walked toward the restaurant, I found myself wanting to be back there very badly. I wanted to apologize for my behavior and make amends to my family whom I love so much. I prayed they would understand and forgive me and the scene of me coming back wouldn't be too awkward. Suddenly, I felt like the biggest brat.

But it wasn't awkward. When I stepped into the restaurant and walked toward our table, I was greeted with nothing but warm smiles. I sat down, and my mom reached across the table to give my hand a love-filled squeeze. Instead of giving me a hard time for acting out the way I had, they showered me in love. "Honey, it is so wonderful what you're doing, and it's an inspiration to see you lead the way," my mother said, still holding my hand tightly. I felt the tears starting to burn again but for a different reason this time. I felt heard and seen. I felt understood.

Then my aunt said something I'll never forget:

You need to understand that you're planting seeds. Every time you walk into a restaurant and order a plant-based option, every time you bring your own utensils, and every time you say or do something you believe in, you're planting seeds. People see that and they begin to process it in their own time. You're touching so many more lives than you know. You've touched my life for a long time already. And before you know it, those seeds will start to grow, so don't ever forget the power you have when you're planting seeds.

I understood at that moment that our power goes so much farther than we can possibly know.

Be a Purple Flower in a Field of Yellow

When you start out on your change journey, you might feel a bit like an oddball at times. You might be the one person who brings a reusable coffee cup to the store when everyone else sips their iced lattes from a plastic to-go cup, or you're the person who wears the same outfit to three parties in a row because you're an outfit-repeat believer and no one else is. Your actions can catch some attention and people might start asking questions. Usually they're filled with curiosity, but there could be times when you might be mocked too. Just remember these comments are coming from their fear, and their subconscious minds feel threatened by your behavior. You're weird, something they're not used to, and their coded egos do not like it.

Don't let it bother you; just be you. Try to have some eye-opening facts up your sleeve that you can pull out for these occasions, things you can say with excitement, and—if you're lucky—plant a little seed. If someone comes to you at a party and says, "Weren't you wearing that the last time I saw you? Don't you have anything else in your closet?" kill them with kindness—*and* with facts. Say, "Yes, you're so right. Thank you for noticing! And you know what? I'm actually trying to slim down my closet and only wearing things I truly love. Have you heard of outfit repeating? It's the new thing, and I kind of love it, to be honest. I'm doing it to lower my footprint on the planet because fashion is actually one of the dirtiest industries in the world, but I'm also realizing that it's so much nicer. I don't have to spend all that time thinking about what to wear—I now have this dress that I love so much."

If you're lucky, you will have sparked a conversation with intrigued questions to follow. If not, ignore it and move on—this conversation is clearly not worth your energy and effort. Don't let any potential comments get to you. Shrug them off and go find yourself something to drink. Know that regardless of the outcome, you've just planted a seed, and you might be surprised by whose colors you're turning....

Think of it as a field of yellow flowers. The only reason the field is yellow is that, so far, there has been nothing but yellow seeds planted there. Humans are tribal creatures, and we want to fit in. It's a primitive behavior, and to be accepted by the tribe, we copy what the tribe seems to be doing. If you live in a field of yellow, it's only because, for the longest time, there has been nothing but yellow seeds.

When you actively choose to be something different—let's say purple—you will stand out a little at first. You will be the one purple flower in a yellow field. However, the more you stick around, the more seeds *you* will start to spread around you. Soon enough, you'll have little purple seedlings helping to plant more of your new seeds. Depending on how sick people are of yellow and how good you are at spreading purple, the field might change colors completely very soon.

Think of your actions the same way. Each time you show up as the change you wish to see, you're planting yet another seed, so keep going.

CHAPTER 61
Do Individual Actions Matter?

One day, Arthur asked me, "Don't you ever get sustainability fatigue?" He was referring to the fact that no matter what I do, I need to figure out the more sustainable option (or if anything, at least explore if there is one). I told him no, and I meant it. In fact, *not* finding a slightly better way of doing things gives me much more anxiety because it means I'm not living in alignment with my values and beliefs. Sure, sometimes you just have to accept you live in a world that is not set up for zero-waste perfection and give yourself a break. However, I will at least try, with the foremost reason being it actually gives me a lot of satisfaction and joy. Figuring out sustainable lifestyle hacks is not a task or burden; it's an adventure! It gives me what I call rebel sanity because at least I know *I'm* doing what I can to make a difference in this world.

Some may argue advocating for individual actions is not the answer. Some even say it's dangerous because it gives people a false idea that they're doing enough. I don't agree. Sure, we must do more than just say no to plastic lids and eat plant-based once a week, but I don't agree that individual actions stop with the individual. I believe they are the building blocks for a whole new system with the potential to reach far greater results than we can possibly know.

I have asked myself about a billion times if my individual actions actually matter. Sometimes I definitely doubt my efficacy. Maybe you've been there too? Have you wondered if, with so many companies continuing to fuel the problem, and with most of the world still clueless about what we need to do, your small actions are worth the effort?

We also have an ongoing debate about who's to blame and who's responsible to "fix this." Is it up to individuals to shop smarter and know better? Is it up to the companies providing our products and services to own up to their responsibilities and change their ways? Or is it up to governments and politicians to use their authority to create new systems and laws that prohibit climate destruction altogether? The answer is yes. We need it all—change has to come from all corners at the same time, which leads me back to the question about your individual actions. Do they matter? Yes, they do.

However, be assured you will still doubt. Every now and then, you will question if your efforts are worth it, and when you do, I want you to come back to this chapter and remember what I'm about to tell you.

Individual actions matter for four main reasons. They:

1. Mend anxiety.
2. Build character.
3. Shift culture.
4. Plant seeds.

If you know the *why* behind these reasons, it will be so much easier (and more fun) to keep going. Let's look at each one briefly.

1. Mend Anxiety

First and foremost, your individual actions help mend anxiety. Since climate anxiety is something we feel when we get closer and closer to an anti-goal (something we don't want to happen), and we feel we lack the power to do something about it, small, individual actions will edge us away from the anxiety and into empowered action. It's about taking back control, and in doing so, mending anxiety.

Call it rebel sanity if you want, and if the only reason you're doing this is to make yourself feel better, then so be it. I think it's totally okay to be a bit selfish in this movement. Also, recognize you can only do as good as you feel, so don't underestimate the power of feeling good about your actions.

Understand that every time you don't agree with something, you have two options. You can either act on it (i.e., do whatever you can, even if it's little) or not. One will make you feel better and more empowered; the other will keep you stuck in hopelessness and despair. The choice is yours.

2. Build Character

Feeling like a hero is awesome, but it's not just about feeling better. Those small actions will start to create new habits and slowly reshape who you are. How you show up in the world and who you identify as will change, and you will bring forth a whole different kind of energy to the world around you.

By being the change, you start to believe in the change, and that is also the energy you're spreading to others. You're proving change is possible and revealing the world you wish to see moving forward.

Moreover, this new identity of yours will make you think differently, act differently, and make different choices. If you carry around your refillable water bottle, walk or take public transportation whenever you can, and say no to plastic in the store, you will think of yourself as someone who cares. Someone who cares is not going to make impulsive choices. Someone who cares is also not going to ignore an opportunity to participate in positive change, be it signing a petition or supporting a local group trying to install curbside composting in your building. Someone who cares will recognize others who care and celebrate their efforts, so by simply being the change you wish to see, you're becoming a powerful and conscious force.

3. Shift Culture

This leads me to reason three, which is how we're continuously affecting the environment around us, whether we know it or not. We are social creatures and, therefore, constantly look to our surroundings—consciously or subconsciously—to see what's acceptable, normal, and cool. The more of us who do something, if it's bringing our own coffee mug to the coffee shop or wearing the same outfit to multiple parties, the sooner that new thing will become acceptable. When something becomes acceptable, it starts to create a new norm, and when that new norm is planted, that's how *culture shifts*. Never underestimate the power you have to shift culture because that is how we start changing the system and, ultimately, the world.

4. Plant Seeds

Last, but not least, remember you're always planting seeds. You may not know how much you're influencing someone, but that person might see what you're doing, think it's really cool, and then try it out for themselves.

The beautiful thing about being the vessel of change is you don't always get to see the results, but rest assured you have made a difference. As long as your actions are filled with light, love, and excitement, as long as you do *you* instead of shaming other people into doing what you think is right, you are planting powerful seeds. Before you know it, you will have a beautiful forest of positive change growing all around you.

Never forget you *are* a vessel for change. You are the hero we need today, and you can choose to activate change whenever you're ready. It's time to recognize what an exciting time we live in, and when it comes to the world of tomorrow, we all have a beautiful part to play, so don't waste yours.

CHAPTER 62
Don't Be an Angry Vegan

One of the best pieces of advice ever given to me came from a bald guy with tattoo sleeves and glasses named Demosthenes Maratos. You may have seen him if you've watched the documentary *Cowspiracy*—he's the guy standing in a field and talking about how much we should learn to love cows.

I met Maratos at a vegan fair in New York City. He was speaking; I was simply attending. He was a veteran vegan with twenty-plus years under his belt. I had just started my journey. Fascinated with everything he said, I wanted to connect, and a couple of months later, we found ourselves chatting over a delicious plant-based meal in Brooklyn. (Tip: If you want to speak to someone you admire, having a platform for an interview or feature, be it a blog or a podcast, is very helpful.)

Somewhere midway through our interview, Maratos put down his fork and asked if he could give me some advice. "Of course," I said. He looked at me like someone about to share something important, and said, "Whatever you do, don't be an angry vegan."

He told me how many of his friends over the years had become what he referred to as "angry vegans." You become so frustrated with the world, he explained, you don't enjoy the journey anymore. You're angry with people who are not choosing the same and your life becomes about either trying to force these people to join your side or ignoring them altogether. "It's an easy way to lose friends," he said, but I could see that without him explaining it. I knew exactly what he meant.

I promised him I wouldn't become an angry vegan, but his words followed me home that night since I realized I was probably on the route to becoming one. When you learn something about the world that upsets you and *you* choose to take action, it's easy to feel like everyone around you should recognize it too and choose to do the same. If they don't, you easily grow frustrated or sometimes even resentful, and suddenly you can only see them through that lens—they're either vegans or not vegans. In other words—either good or bad.

Not only will this ruin some valuable relationships, but you're actually ruining the journey for yourself too. Your choice to become the change soon turns into this battle where every meal and every interaction serves as an opportunity to "do your duty"—to convert someone else to doing the right thing.

To state the obvious, being an angry vegan isn't fun, but it also doesn't help your mission. When you come at someone with shame and blame, that person will only feel attacked and find a reason to defend themselves. Wouldn't you? Besides, if all they see of "vegans" is how angry and frustrated you seem to be all the time, you're not showing good reasons for considering being one.

Instead, if you skip the angry part (I know it's hard, but practice will help), you can keep doing what you believe in with light and excitement. Order the food you want without comment. Bring your own snack to the party if you want to play it safe, but don't make a fuss about it. Continue to choose plant-based options, and do it because *you* chose this way of life, but without feeling the need to convince anyone around you. If you stay quiet and let your actions (and good energy) speak, you'll convince more people than you're aware of.

That is how energy spreads. It's also how we share new norms and shift culture. If you order a plant-based option off the menu, you make it okay for others around the table to order it too. If you suggest to your friends they should check out the new vegan shack in town (not because they *morally should*, but because the food looks delicious), you're giving them a chance to try it out on their own. By providing opportunities for the people around you to try something new by themselves, they get to have their own free experience, and the chances of them associating it with something positive are so much higher. That is why you want to be a *happy* vegan—someone who shows how *awesome* life can be this way, and in doing so, invites others to join you on that journey. Lead, don't push, and be the reason you believe in better choices. If you believe it, chances are others will start to believe it too.

Needless to say, this applies to anything you're passionate about, whether it be zero-waste living, solar panels, or skipping the straw. And it goes for non-environmental missions too. Whether it's health routines, philosophies, or newfound habits—whatever it is you believe in—you have a much bigger chance of inviting people in *and* enjoying the ride if you act from a place of curiosity and joy. Good energies are addicting. It's as simple as that, so remember that when you're out there trying to change the world.

I also want to admit that although I was strictly vegan for five years, I no longer call myself a vegan. Becoming vegan was essential for me to start my conscious food journey, but as the years went by, my awareness grew, and I learned nothing is black and white. In fact, vegan does not necessarily mean ethical, sustainable, or even cruelty-free—it simply means no animals were used in either the process of creating the product or in the product itself. However, if the people picking berries do so under inhumane conditions or if the vegan product is created with toxic chemicals that disrupt ecosystems and kill wildlife, then "cruelty-free" doesn't really apply anymore. I'm not saying that vegan isn't a better option, because in most cases it is. However, by looking deeper at our choices and by constantly being curious to learn more, we can expand our horizons and keep striving for even better values and ideals. (More on this in the next chapter.)

CHAPTER 63
Titles

I used to love and live by titles. I found them so helpful in setting boundaries for myself and who I wanted to be. I was repulsed by what I learned about the industrial farming industry and wanted nothing to do with it moving forward. Shocked to learn I had contributed to this kind of pain and suffering my entire life, I needed to set rules to ensure this would not be the case anymore, and being vegan rescued me.

All of a sudden, it was so easy to do the right thing. All I had I do was always choose the vegan option—no matter what—and if there wasn't one, simply skip the meal altogether. (I ruined the first three days of our honeymoon in Italy trying to avoid all things non-vegan and gluten. Let me tell you, my body soon got sick of olives, nuts, and prunes.)

However, after a few years of rigorously living this way, I began to question again. To stay within my boundaries and keep my spirit and consciousness clean, I had ignored information suggesting anything that wasn't aligned with my already existing beliefs. When people spoke of the importance of animal grazing for soil health, I wouldn't hear it, and when anyone dared to say that vegan wasn't always the best option, I crawled further into my mentally made-up walls where I figuratively plugged my fingers into my ears. *La la la la. I don't want to hear it!*

Childish, I know, but this is honestly how most of us live. We tend to defend ourselves and our behavior by consciously and subconsciously filtering out information that doesn't align with our current beliefs. Think about this the next time a family dinner turns into a heated argument and everyone seems to be going to great lengths to prove they're right, with more and more absurd arguments backing their point of view.

At some point in my vegan journey, I found the courage to let my guard down and explore new options. I recognized my openness to change was what had led me to this awakening in the first place, so why would I close myself to that source now? Being vegan had served me, but was it the best way forward?

I decided to dig deeper into areas that intrigued me, like the world of soil health, biodiversity, and regeneration. With the books I'd read and films like *Kiss the Ground*, *The Need to Grow*, and *Sacred Cow* under my belt, I began to see there was more to it than just cutting out all animal products. Things like crop rotation, animal grazing, and protecting native species played a huge part in restoring Earth as well. I also understood being a vegan didn't remove me from all the death and harm inherent in the system. Death is part of the cycle of life after all, and even if I eat only vegetables, rest assured someone had to die in the process (even if those were the slugs killed to save your sacred lettuce).

The truth is, as long as we're simply consumers and go to the grocery store for our food, we will always remain somewhat removed from the reality of the system, but that doesn't mean we can't become engaged and learn. Read up on regenerative agriculture (or sign up for many of the classes offered online), hit up your local farmers market, talk to farmers, and stay open-minded and curious. Accept how you're living and eating now may not resonate with who you want to be, but tell yourself it's okay to reach this realization, and then know you're allowed to choose again.

I began ditching titles for this very reason. When you put yourself in a box like that, it becomes very hard to move away from your original decision. If you're a vegan and you go out to eat with friends, for example, it's suddenly hard to order anything but the vegan option. But what if that one option is an exotic dish made from foreign foods shipped in from miles away, while all the other options are locally sourced from nearby farms with free grazing, happy animals? Suddenly, your option may *not* be the best one, but rather the one that will most likely fuel climate change and contribute to eco-system disruption (with shipment like air travel polluting the air and big marine transportation interrupting ocean life), compared to the local ingredients that are empowering neighboring farmers and helping them become positive stewards of the land. Many times, if not most times, a locally produced crop (especially if organically and regeneratively produced) will be the absolutely most sustainable option.

This very realization hit me when I moved to The Berkshires in western Massachusetts during the first two years of COVID. Driving around, I could see the cows and chickens running around on lush fields, and it felt strange *not* to eat the scrambled eggs my hubby cooked for breakfast. I allowed myself to recognize that feeling and tuned into the deeper knowledge it carried. Was I holding myself back? Was I missing something very important? I knew then it was once again time to make another choice.

Titles can be good, but they also box us into different categories and, in many ways, limit who we are and who we can become. Today, I try to skip titles as much as I can. I'm not a vegan, but I eat plant-based options most of the time since it's still the food that makes me feel the best, and I won't sway on my ethical beliefs unless I know the source of the animal protein and how it was raised.

Although I was born and raised in Sweden, I almost feel more American than Swede these days, so I guess I'm a human being from Sweden who currently resides in the United States. I live in New York, but I'm not sure I think of myself as a New Yorker because I love the countryside just as much. I guess I'm simply a wild being in love with many things.

What I've learned is nothing is black and white, and things are always more complex than they look on the surface. Without titles, you empower yourself, and every choice becomes a conscious one. It's not so much about who you are, but what you *do*, which allows you to address each situation with intention and thought.

Besides, if you skip titles, you're much less likely to exclude people who don't identify the same way, and you allow for inclusivity and deeper conversations.

"Oh, I wouldn't call myself a vegan, but I love plant-based foods! Besides, it's so much better for the environment. Did you know…?"

By approaching the conversation this way, you are telling the other person they won't get stuck if they try it. They can always choose again. When people feel they are in control of their options, they are much more likely to take action.

By skipping titles, you:

1. Give yourself the power to choose what's best for any given moment.
2. Allow for self-reflection, new knowledge, and growth.
3. Appeal to curiosity when you invite others so it might spark positive actions on their end as well.

Note that not all titles are bad, nor are they all disempowering. Sometimes titles are useful for finding an identity. Sometimes we want these social boxes because they help make life a little easier. For someone struggling with drinking problems, for example, finally saying out loud, "I am an alcoholic," can be the critical turning point in their survival. In those cases, claiming a title can make or break your life.

Therefore, I'm not saying you should skip titles altogether; simply start by reflecting on the choices you're making based on your different titles and check in with yourself—are you holding yourself back? Could you, in fact, empower yourself if you skipped one or two titles?

Too often, we create our own limitations with the words we tell ourselves, and we miss that we actually have the option to choose or not choose that same reality for ourselves moving forward. Ask yourself if you want to be who you are now forever, or do you want to grow and expand? If the answer is to grow, then start *doing you* instead of *being you*. The doing can always change, and your being with it.

CHAPTER 64
Successful Climate Communications

Apparently, only one in four Americans hear someone they know talk about climate change every month. Considering it's a threat multiplier for all the things we care deeply about—health, safety, economy, and social justice—it's crazy to think that 75 percent don't talk or hear about it even monthly. The reason is understandable. Since climate change is perceived as an overwhelming threat, the everyday person feels powerless to do anything about, so why would we want to talk about it? It is much easier to be oblivious and continue with our lives as usual.

Another reason we don't talk about it more often, as explored earlier in this book, is most of us perceive climate change as distant either in time or space. Since it's not a threat facing us here and now, focus easily turns to gossip, TikTok trends, and other important-but-also-not-so-important matters. We're humans after all, and we live in a world that's always asking for our attention, so how can we put our minds to something so complex, overwhelming, and scary?

The truth is, as you know by now, climate change is already here and changing all our lives to one degree or another. Climate change is not distant in any way, shape, or form despite how we may perceive it. Therefore, it's critical we start giving it the attention it deserves so we can give ourselves a chance to act before it is too late. That starts with conversations. The more we can normalize talking about climate change, the faster we can spread awareness and invite more and more people to shift toward a better world.

"The most powerful thing we can do for climate is talk about it."
— *Climate Scientist Katharine Hayhoe*

However, having successful climate conversations can be tricky. An already loaded topic, it's easy to find yourself in a frustrating situation where your counterpart either doesn't want to hear what you have to say, doesn't believe what you're saying, or argues something completely different is true. This is normal. Climate change, to many, serves as a shame trigger. Subconsciously, we see it as a reminder of how bad we are and how we are to blame for the damage we're creating.

We know this is only partly true and that it is important we shift the narrative on that story because if we continue living in shame and denial, we won't take action. But how do you start conversations that spread awareness and invite action, instead of growing frustration and shutting doors? It will take practice, but here are five tips for the next time the opportunity presents itself.

1. Lead With Questions and Find Common Ground

When you want to really push a message, it's easy to come off as a bit too…what's the word…excited. You feel like you've learned something the whole world needs to know and you forget the people you're talking to have yet to share your excitement (or worry). This can backfire. If you start a conversation by trying to convince someone what you're saying is important, you run the risk of losing them early on.

When it comes to climate change, things get even trickier. For this topic, we carry a lot of pre-established fear, shame, and denial. We're used to associating climate change with an attack on our personal lifestyles and behaviors and, if you're not careful, you can launch the receiver right into a predisposition defense.

To avoid this, lead with questions. That way, you open the conversation to curiosity and an even playing field. Ask what they think about (wildfires, flooding, the fact that your local river is getting polluted, or something else climate-related) and genuinely listen to their responses. Pay attention to what they say and try to find common ground—something you both care about—to build off of. If the person feels listened to and understood, and if they recognize you care about the same things, they will be much more likely to hear what you have to say.

2. Talk Local

Just as finding common ground is an important foundation for successfully engaging someone, so is talking about something the person can relate to. In climate conversations, this is even more important than you might think.

For most of us, climate change doesn't feel like a personal threat. Sure, we all think it's sad that polar bears are starving and people in other parts of the world are losing their homes. Of course, we *care*. But it doesn't matter how much we care; to *activate* someone, the problem needs to feel relevant and close to home.

Remember that one of Per Espen Stoknes's 5 Ds (barriers for climate action) is distance. That means we feel like climate change is both geographically and socially distant. It will happen in the future, not right now. It affects people in other countries, not us. This perceived distance prohibits the psychological urgency needed to take action.

By talking about how climate change is affecting your local area, you are much more likely to instill interest in the people around you. And (un)fortunately, finding examples of how our warming climate is affecting life where you live is not hard. All it takes is a few minutes on Google and you have yourself a conversation starter.

3. Don't Shame!

Maybe it goes without saying, but shaming people is not going to make them care more. Yet that's how we often go about educating people on climate change.

"Don't you know that driving is terrible for the planet?"

"How can you buy yogurt when it comes in plastic?"

"You're gonna order a steak, really? Don't you know anything about Amazon deforestation?"

You know as well as I do that these are all solid claims, but they have little chance of actually converting someone into a climate ally. If you attack someone, they usually don't embrace your attack. Instead, they go into defensive mode and shut you out. They don't hear anything else you say. Sure, you might get your point across and get noticed as an obnoxious "eco-hero," but rest assured, you're not changing hearts by acting this way.

Try to avoid framing that might lead to denial or the need to protect oneself. If you tell someone off by pointing out how bad their behavior is for the environment, chances are they'll only look for excuses as to why changing their ways won't matter (and even if they don't say it to your face, they surely will argue it to themselves later on).

Instead, you can say something like: "You know, I learned one truckload of plastic is dumped into the ocean every minute; isn't that crazy? I decided to get a reusable water bottle, and I feel like a hero every day now, not having to throw a disposable bottle out. Would you like to try getting one too?"

This way, you share information about the issue without attacking the person you're talking to. Instead, you're putting yourself center stage and sharing how much *you've* changed since learning these facts. Then you invite the person to join in on that change.

4. Use Storytelling Over Numbers

As humans, we're natural storytellers. That's how we've been passing on ideas and values since we learned to talk. Before we had smartphones, movies, and social media, we had books, and before then, we had campfires and the spoken word. We gather around stories and have for as long as humans have had language, simply because stories *move* people. Also, many if not most of our decisions are emotion-based, so if you want to spark action in someone, aim for the heart. Stories get to the heart faster than facts and figures.

Loading someone with facts and figures might feel like you're building a powerful case, but remember it feels that way for you because *your* emotional side has already been activated. Once you care passionately about a subject, finding facts to back your beliefs will make it seem like a bulletproof case. However, for someone who's yet to "get it," facts and figures are hard to relate to. You can look at a graph and understand the scale of the problem, but that doesn't mean you're emotionally triggered to take action. A lone polar bear on a floating ice sheet or a homeless child forced away by flooding will have a much stronger effect. That is why you want to find ways to focus your message on people, not numbers, and activate that caring heart.

Instead of talking PPMs (parts per million in the atmosphere, as in how we measure CO_2 emissions) and GHGs (greenhouse gases), find a story about someone affected by climate change (preferably someone local to the person you're talking to), and use that story to start a conversation. Or if you feel like it might be more inspiring, talk about how climate change will hurt *you*. What will your community and country look like twenty years from now if climate change continues to run its course? What will it be like when the local store doesn't have fresh food? Even if you can't find information online, use your imagination—what do you fear will happen if we don't act? Share that fear. Once someone is genuinely engaged in the topic, you can use the numbers to hone your point.

5. Talk About Solutions

More awareness does not spark more action. We already covered this, so it is nothing new. However, when we talk about climate change, we tend to do just that—we keep spreading awareness of the issue with very little information about what to actually do. If this is how most people hear about climate change, it's no surprise we don't want to talk about it more often.

That brings me to the obvious—we need to talk about *solutions*—and thankfully, there are many of them. The last section of this book includes some climate solutions to be aware of and get excited about. I recommend getting involved in one or a few that really speak to you.

Something that became clear to me when I interviewed Katharine Hayhoe for the *Hey Change Podcast*, something I hadn't thought of before, is that often the climate change solution is the cherry on top, an added bonus that comes with simply making better choices in the first place. She said it with so much joy and excitement, and it made so much sense. Here's what she said:

Indent next paragraph as quote

By planting trees in urban areas, we make the environment prettier for those who live there; the summers will get cooler from the shading trees, microclimates will stabilize, and the air will be so much nicer to breathe—and you're helping fight climate change.

Katharine gave example after example of how this is the case, from switching to solar panels to driving an electric car.

The truth with many if not all climate solutions is they are cheaper and better options for so many aspects of our lives. We just have to be curious enough about the change to see it. When you talk about climate change and invite people into solutions, don't sell the idea of a moral obligation—get them really excited about all the benefits instead.

Infusing hope and excitement into climate conversations is obviously essential, but there's an additional psychological benefit from sharing positive solutions that already exist. When you let someone in on the positive already taking place, you let them jump on board and be part of the journey, which automatically triggers the brain to want to go for it. If we feel like we're already one step on the way, it's much easier to get inspired to take action.

Having successful climate conversations takes practice, but as with everything, it will get easier the more you do it. If someone starts attacking you, don't take it personally. Climate change is a very loaded topic and can trigger all sorts of things; fear, denial, shame, despair. Yet, we need to have these conversations, so don't let that scare you—be the hero and keep spreading both positivity and awareness. By using the steps above, you should hopefully be able to steer the chat in a more empowering direction.

A Few Extra Tips

- Sometimes it's better to avoid using the term "climate change," especially with people who tend to deny its science. Instead, look for something you can agree on. Maybe the fact that the weather is changing (i.e., getting weirder) is a good starting point.
- Don't overwhelm the person with too much at once. When you get a conversation going, it's easy to get excited and want to tell people everything you know. Remember it took a long time for you to get to where you are today, so tread lightly. Focus on one climate topic right now and build from there.
- If you fail to spark a conversation, don't beat yourself up. You probably planted a seed even if you don't know it.
- You don't have to have all the answers to spark a conversation on climate. In fact, coming at it with humility and transparency invites

others to feel like they can partake too, even if they're not climate experts. Remember, simply starting conversations is the goal.
- Have fun with it, and treat it as an empowering journey. What can *you* learn from starting these talks? Stay curious and open.

CHAPTER 65
Lagom—Not Necessarily Perfect

I love the Swedish word *lagom*, and I'll probably never stop using it no matter how long I live abroad. It basically means to do things "good enough, yet sort of perfect." *Lagom* can be used for anything from how much milk you'd like in your coffee to explaining the weather or describing how big of a house you want when you grow up. It's not too much or too little, not too hot or too cold, not too small or too big; it's somewhere right in the middle—it's "just enough." It's not the same as perfect, but a bit mellower than that. It's *lagom*.

It might seem like a boring concept, but don't be mistaken—Swedes love *lagom*. *Lagom* is good, a life standard to aspire to.

"How much milk would you like in your coffee?"

"Oh, just *lagom*."

Of course, *lagom* can be very different for different people, so you expect the other person to indicate when you've poured the *lagom* amount.

"Thanks. That's great."

What I love about *lagom* is that when you google it, you'll find the definition "not necessarily perfect." You understand that the outcome may not be perfect, but it's okay because *lagom* is often more than good enough. When it comes to our climate actions, that is exactly the kind of mindset we need.

We must recognize questioning the *status quo* and slowly moving in a new direction will come with our fair share of mistakes. If we strive for perfection, we will fail, and grow more and more frustrated in the process. However, if we aim to do as good as we can while also ensuring we tend to ourselves—if we infuse a little bit of *lagom* into our every action—I believe we can be successful. Slow and steady wins the race, as they say, and I think nothing could be more true about our climate actions.

Don't mistake this for being lazy or for not caring. *Lagom* has everything to do with reaching the outcome you want—you're simply being more open-minded about what that will be. We must certainly strive for the best there is and not sway in our vision of the world we want to see, but as we embark on the journey, it's important we don't get stuck seeking perfection.

We also need to act now because we don't have a second to lose, so this has nothing to do with being slow, but the act itself needs to come from a place of grounded intention. We have to remember the road trip and continue to find ways to fuel ourselves *and* the action, and keep doing what we can, step by step by step, to let go of the old and head toward even better worlds.

Not necessarily perfect also speaks to finding the courage to try new things. It's not going to be perfect right away, but as you try, you learn, and you will get better and better as you try again. Don't fear acting because you're not a climate expert—few of us are. Recognize you, too, belong on the climate hero squad, and whatever you can do today, is absolutely enough. It's *lagom* for you, and as you keep growing on this journey, that *lagom* will change, and you will keep raising the bar of your own eco-hero potential.

CHAPTER 66
Slow

Did you know an average American college student switches tasks every sixty-five seconds? Or the medium time they focus on any one task is just nineteen seconds? How any papers get written these days, I don't know, but it seems like a miracle people are even graduating. However, it's not just the young who are lost in distractions—the average person touches their phone 2,617 times every twenty-four hours. I didn't know we had time to do *anything* 2,617 times in a day.[33]

Our focus is being stolen from us, and if this worries you even slightly, I recommend checking out Johann Hari's book, *Stolen Focus*. However, I want to focus on what this is doing to our ability to trust the path we're on.

If you are constantly swept away by distractions, stimulation, and other people's opinions, how do you know you're on track with what *you* want, both for yourself and the world? How do you know if you're living your truth or someone else's?

In *The Influential Mind*, Tali Sharot goes into detail about just how impressionable we are. She shares in her book that studies on our brains' amygdala (the "old" brain where feelings like fear are stored) indicated we are so influenced by other people's thoughts and opinions that our own memory can be altered forever. She writes: "When your amygdala reacts very strongly to other people's opinions, it triggers a biological reaction that prevents your frontal lobes from subsequently correcting false beliefs."

Obviously, the world around us shapes us, but if we don't pay attention, we can soon head down a path we don't recognize as our own.

---

[33] Naftulin, Julia. "Here's how many times we touch our phones every day." *Business Insider*. July 13, 2016. https://www.businessinsider.com/dscout-research-people-touch-cell-phones-2617-times-a-day-2016-7

We pick up new trends and hobbies because we believe we should, rarely paying attention to what our own hearts truly desire.

Learning something new from others is often a good thing. This influence gives us the power to help others and shift norms and cultures for the better. However, by recognizing just how impressionable we are and how much time we lose to distractions we're barely aware of, we can begin to understand how much we actually need to take a break and find ways to slow down.

Slow has gained a whole new meaning to me in the past few years. I've realized that slow is a mix of luxury and strength—luxury to have the time to slow down, but also strength because it takes a lot of hard work to get to the emotional level to allow ourselves to slow down. I think for most of us, slow is a state we crave, but once we get there, it feels so uncomfortable we quickly go back.

When we think about "climate action," we tend to think we have to do, do, do—more, faster, *now*! However, a big part of that "doing" is also about doing less, and it's critical we begin to see this. When we slow down and quiet our minds, we begin to recognize many of the answers we're looking for are right there, already nested inside us. When we slow down, we create time for our minds to wander, and we birth opportunities for new creative ideas to appear, ideas that could be the next great solution we need to move forward.

When you slow down and quiet your mind, you have time to remember who you are and the things you actually care about, which most likely will be all the things you already have—time with those you love, a quiet cup of coffee in the mornings, good dinners with friends, or a long conversation over a walk in the park.

More often than not, the most valuable things are the simple things you could enjoy more right now if you only slowed down and paid more attention. Allow yourself to get in tune with life and your own body, and in doing so, realize you already have so much more than you could ever want.

When we start to appreciate the small moments, we automatically move away from things that stress both humans and the Earth, like consumerism, traveling just for traveling's sake (do you really need to leave home every vacation?), and the constant need to have more. I always marvel at how at peace I can feel on a summer morning in the country, in a mismatched outfit and bare feet on the grass. In those moments, I get to experience what it truly feels like to be alive, and I know in that moment I don't need anything else to fulfill me. I feel happy, grounded, and at peace with myself and the world. I always promise to remind myself to find more of this in my otherwise hectic life. Because if we could only do more doing less, so many of our problems could be solved.

It can feel uncomfortable to slow down sometimes, especially if you're used to always being in the middle of the action. You might even feel guilty for "taking a break" when the world needs you so much. Remember this, too, is an action, and by slowing down, you're anchoring nourishing energy in the collective consciousness and (unknowingly) inviting others to slow down as well.

We don't need a world full of panicked chickens; we need graceful, wise owls taking in the entire field and making decisions based on what seems best. Be the owl, not the chicken, and continue to ground yourself as frequently as you can.

A Slowing Down Challenge

If you like a challenge, here are three kinds of slowing down you can practice this week to help you recognize what "slow" might look like and how easy it is to actually invite it in.

1. **Active Slow Time:** Schedule at least one session of active slow time when you allow yourself to do nothing at all. This doesn't have to be long. Ten to fifteen minutes is enough. The important part here is allowing all of yourself to slow down—your brain as well. That means no reading, tweeting, scrolling on social media, watching TV, and preferably, no talking. This is a time just for you, with zero distractions. Can you spend fifteen minutes just being?

2. **Integrated Slow Time:** You actually don't have to set aside time to slow down, and once you get better at it, you'll learn it's a mindset more than anything else. By learning how to slow down in your everyday life (yes, this is possible for busy city dwellers too), you

will notice it's so much easier to be present. In the present, joy and abundance flow. For example, if you're walking around the grocery store, try to slow down and pay attention to your experience. What do you actually want to buy? What do you tend to buy out of habit? What new brands could you try that might align better with your current values? By slowing down in daily life, you not only bring more joy to all the things you do; you also unlock the ability to think again, stay curious, and explore new stuff.

3. **Slowing Your Thoughts Down:** We don't usually recognize how fast our brains race, but oh, our thoughts tend to take off, don't they? We read something that triggers us, someone says something that upsets us, or we fail at a task and spiral into the "I'm not good enough" trap. If we can slow down our thinking, we can recognize our experiences for what they really are: lessons and growth opportunities. When you sense your thoughts and emotions racing off, take a deep breath and slow yourself down. How can you look at the situation with different eyes? What hidden blessings and lessons are you not seeing? When you slow down your thoughts and feelings, you put yourself back in control, and activate a way of being that enables self-love, empowerment, and growth.

CHAPTER 67
I Think You Should Hug Trees

I'm a proud and avid tree-hugger. Not just a metaphorical one, but the kind who actually goes out there, puts her arms around a tree, and gives it a squeeze. After the squeezing, I like to lean my cheek or forehead against the tree's rugged skin and connect for a moment. I want the tree to feel I'm there just as much as I can feel it, to let it know we're out here doing what we can to protect it. I hope the tree understands my message and uses its access to the mycelium—this magical web of life we cannot see—to pass it on to the trees around it so they can pass it on to the trees beyond them, and they the trees beyond them, and so forth. I want the natural world to know we're not a lost cause, that, in fact, humanity is fighting to find a way back and learn how to respect life the way we ought to respect it. I want them to know we've not given up.

Yes, I'm totally the nerd who talks to trees, and I highly recommend joining me on that squad. Hugging trees and touching nature is beneficial in so many ways. Hugging a tree increases oxytocin, the amazing hormone responsible for emotional bonding, and serotonin and dopamine, which make you feel calmer and happier. It's also been proven that "eco fitness" (spending time in the great outdoors) helps you stay healthier, happier, and better able to focus at work.[34]

Personally, I use tree-hugging for daily grounding, and it works as well in the city as it does somewhere deep in the woods. I may have worried about people wondering what the heck I was doing for a while, but once I got over that social hurdle, I began hugging city trees too. I think they might need it more than all the others.

When I hug a tree, I place my feet solidly on the ground so I can feel

---

[34] Weir, Kirsten. "Psychological research is advancing our understanding of how time in nature can improve our mental health and sharpen our cognition." *American Psychology Association*. 51.3. April 1, 2020.

the energy from the earth beneath me. I let that energy travel up my legs as I connect with the energy inside the tree. It might sound hokey that I can feel it, but I can—this energy is so palpable that all you have to do is take a second and tune in. Plus, trees are always welcoming, so when you really need a comforting hug, you know where to go.

If you've never hugged a tree, I challenge you to try it. Find a secluded tree without an audience so you don't have to worry about looking like a fool while you do it—it's important to let go of ego and simply tune in.

If a big old squeeze is too much for you, simply touching nature can do the trick. Put your hand on the bark and connect that way, or touch some moss, a rock, or sand if that's what you have around. Also, try to be barefoot in nature as much as possible, as this too has proven to have significant benefits for both our physical and mental health.[35]

If you have hugged a tree before, I challenge you to do it more often, and to find your arms around those beautiful city park trees too. How wonderful would it be if suddenly people all over were hugging trees on their daily commute and good energy was spreading from the branches? Imagine the world *that* would be. I can just see the headlines: "New trend of tree-hugging is spreading across the land, with happier employees as an unexpected result."

Sometimes, the solutions are as simple as hugging a tree, and you can decide right now you have the courage to do so. Life's too short—go hug trees!

---

[35] "Feel the Earth Beneath Your Feet—the Benefits of Barefoot Walking." https://combegrove.com/health-and-wellbeing/2021/the-benefits-of-barefoot-walking/

CHAPTER 68
How to Think Like a Climate Optimist

Mindset is everything, even when it comes to changing the world. However, staying in the right one can be hard, especially with so much stigma and fear-mongering circling around out there. Here are some mindset shifters to help with remaining not just optimistic and empowered, but open-minded and curious. Simple shifts in how we think about ourselves, the world, and our actions can make a critical difference in succeeding moving forward, so don't underestimate the power in your own words.

Read the following mindset shifters to see if they resonate with you.[36] Are you the *status quo* or the climate optimist? If you lean toward the first one, how can you embody the climate optimist within?

Remember, to be a climate optimist is not to sit on the sideline and hope for things to turn out for the better, but to show up and do the work. Use these mindset shifters to practice a new mindset, and see how they might shift your life and even the world around you. Did you think it was impossible? Well, think again.

Come back to these when you need to remind yourself of who you want to be, the world you want to believe in, and the power you have to make a difference in the world. If you want, choose a favorite and hang it on your wall. Then start every day according to this new mindset. A climate optimist is ready to be born.

Status quo:
"It will never happen. There is no way people will be ready to [blank]."
Climate Optimist:
"What current system/belief is behind this mindset? What is blocking our ability to see opportunities for growth?"

---

[36] You can find downloadable versions of these mindset shifters on theclimateoptimist.com.

Status quo:
"The problems are too big, and I'm too small to make a difference."
Climate Optimist:
"What *one* thing can I do today/right now that will make a positive difference? How can I start?"

Status quo:
"They (other countries/the government/companies/other people) have to change."
Climate Optimist:
"What is happening in *my* community? How can I help drive change locally?"

Status quo:
"This is the only way. To solve climate change, we have to do this."
Climate Optimist:
"Why is that? And what could we possibly be missing by thinking this way?

Status quo:
"Climate change is too overwhelming to think about; it's too much."
Climate Optimist:
"Let's start talking about climate change so we can normalize these conversations. I don't have to be a climate expert to have a say in this."

Status quo:
"The climate crisis is too important; all other issues have to wait."
Climate Optimist:
"Yes, climate change is a threat multiplier, but climate justice cannot be achieved if we don't also work for social justice, equity, public safety, biodiversity, and human health. However, this work goes hand in hand—working on all the issues at the same time will help us achieve each goal faster."

Status quo:
"It's already too late."
Climate Optimist:
"We have nothing to lose and everything to win. We can do so much to reverse global warming and build a sustainable and equitable world (science proves it), so let's act now!

Status quo:
"I can't make a difference."
Climate Optimist:
"Whatever I do, I'm planting seeds. By being the change I wish to see in the world, I'm part of the climate revolution."

Status quo:
"To acknowledge and accept climate change for what it truly is means to live in fear, shame, and regret."
Climate Optimist:
"As a climate optimist, I understand it's my duty to live life to its fullest and do the best I can to bring happiness, peace, and love into the world. That is the only way I can actually make things better."

CHAPTER 69
Leaders Create Leaders

When I heard this quote in a Climate Reality workshop
with Al Gore, something clicked deep inside me:
"Leaders don't create followers; they create more leaders."
— Tom Peters

I understood then what our mission really is and how we are going to make it out of this climate crisis. It's not about assigning one leader to point us all in the right direction, but about awakening that leader in every single one of us. It's about trusting our inner knowing. If we can just awaken our light, new truths and wisdom will be spoken. I might be bold in saying I believe in humans and our ability to make this right, but I inherently do. It may not happen all at once, but as more of us start to build bridges and show others the opportunity for a different way of life—a different way of the world—new leaders will awaken in all sorts of places, doing what they can in their small communities and homes.

One leader can't be everywhere at once, but if we have people acting like leaders all over, there's no need for that.

Besides, you know your community best. You understand your family and your friends, and you know the kind of world *you* want to see moving forward. That is why you can be the leader in your own life like no one else can. That is why someone from the outside will never be able to mobilize your people like you can. You have the power to become a leader, in your own life and in others'. I truly hope that after reading this book, that is exactly what you want to be.

The quote says to truly be a leader is to strive to create more leaders, not followers. This means you're not out to make other people do exactly what you say; your purpose is to awaken in them the same curiosity you feel so they will explore these options on their own. When they get to reinvent themselves in this way and step into an empowered position leading toward positive change, a new leader is born, and that leader will soon be out awakening even more leaders. That is your mission—to be the most grounded and empowered leader you can be, and in doing so, awaken new leaders all around you, leaders who are not afraid to question, who understand the power in their actions and words, and who want to see a better world too.

When you start this ripple effect of change, you realize it's not just about you, but about all of us, and that through this ripple effect is exactly how we will make this happen.

We won't survive the climate crisis because some superhero finally decides to save Earth, but because we will. We will get through this and out on the other side into a world that may be much more beautiful than anything we've seen today because of the leaders who will be born inside us.

You can activate this chain of change today. All you have to do is remember what you've learned in this book and set empowered action in motion. Nurture yourself, see yourself—become the leader you were put on Earth to be. Activate the leader within, and as you keep your fire burning, watch in amazement as the people around you start doing the same.

Now we're changing the world for real!

CHAPTER 70
Scary AF

Sometimes things get scary AF (as fuck). You feel like the world is closing in on you, and it gets hard to breathe. At least I feel like this sometimes, and I ask myself why the hell I'm even trying.

Why do I try to make things better? Why do I believe? The world is clearly headed in the wrong direction and there's no way in burning hell that I—tiny little me—will be able to change that.

I feel like an imposter, a fake, a joke worth laughing at. Only I don't want to laugh. I want to cry. I want to scream from the bottom of my lungs and silence myself forever, all at the same time.

Sometimes I feel like I want to hurt myself because the pain would be easier to take than this, less difficult to deal with than the bottomless despair of not knowing.

But then, something turns. Something deep inside stretches out a hand and says, "Stop it. It's all in your head. The stories you tell yourself right now, they're not real."

Are they, or are they not?

I am hesitant enough to listen, unsure enough to seek more. I want to know the answers. I want to find the curiosity to explore again.

What if there *is* something better? What if we *are* headed in the right direction? What if I *am* making a difference? Maybe all I have to do is take a deep breath and…just be—be in the chaos. Exist in the stillness. Believe no matter how messy things look on the surface, a dance is happening somewhere, and I need to find my way back to that dance, immerse myself in the rhythm of change, and be one with the chaos and wonders.

Inside the dance, there's no right or wrong, only play. Inside the dance, there's no space for shame, blame, or hate—I'm too busy keeping up with the steps.

I want to be immersed in that dance forever, curiously seeking the next step, but I do have days when I'm tossed to the sideline, and all I can see is the chaos appearing before my eyes. On those days, I'm exhausted, but I allow myself to have those days too. I tell myself it's okay to live with a bit of worry, anxiety, and fear, because those, too, are credible feelings.

But then I dust myself off and move back into the crowd because I know I'm not done dancing. Life is a dance and a journey I'm not willing to retire from, so I will keep at it as long as I live. My moves may change over the years, but I'm pretty sure there will always be room for me somewhere on the floor.

Because I'm not done living. And I'm not done seeking. And I will never be done believing that there are more, deeper, and better things out there worth looking for.

Allow yourself to have those days—the days when things are scary AF. But then dust yourself off and get back to the dance floor because we need you in here with the chaos!

CHAPTER 71
Things Will Get Worse

Imagine working hard to achieve positive results but the opposite is happening. Not only do your efforts *not* appear to be making progress, but you're actually going backward. Things are not getting better, but are, despite your efforts, continuing to get worse.

I want you to imagine this—*think hard*—because it is exactly what it has, is, and will be like for a while with our climate work. Enough carbon dioxide and other greenhouse gases have already been spewed into the atmosphere to trigger a chain of negative effects. In scientific terms, we've reached tipping points, critical thresholds that, when exceeded, lead to large and often irreversible changes in the state of the ecosystem. The scariest thing about tipping points is that, although we have an idea of what might happen, we actually don't know. What will happen when all Greenland's ice sheets are gone? How much exactly will the sea levels change, and what will that do to people, infrastructure, and other species? Melting ice sheets is just one tipping point; we're racing toward many more.

We are disrupting the world's natural ecosystems and have been for way too long. It will continue to lead to even weirder weather, more intense storms, and more unpredictable seasons. As a result, species will continue to go extinct, people will have to flee their homes, political turbulence will worsen, and everything from food to energy prices will go up. It's the reality we created—the bed we made for ourselves—so it is what it is.

Unfortunately, CO_2 doesn't just disappear once we stop producing it, so even if we were to quit burning fossil fuels tomorrow, there's already enough in the atmosphere to keep the globe warming for a while. We've "baked ourselves in," so to speak, and we will continue to deal with this old damage.

I want you to understand this because even if we *are* heading in the right direction, things will get worse for a while. Knowing this is critical because we must be able to witness the chaos and still not give up. We must continue to have faith and know that we are making progress, that we are making the right choices, and we are building toward a better and more sustainable world. We must continue to ramp up our efforts and do all we can to halt climate change and remember that, yes, we are making a difference, and even though we can't see the results yet, we are headed in the right direction.

It will be tough. Watching the world around us get more and more unstable is going to take a toll on us, which is why the work shared in this book is so important. I want you to work on emotional resilience and heal yourself regularly so you can continue to show up and be a powerful force in this work. I want you to understand how important it is to dream, believe in even better realities, and use that as your motivation to fuel your efforts. I want you to practice humility and open-mindedness, to continue to ask questions and seek even better answers. I want you to learn from others, share what you know, and find community in everything you do. If you do this, if we all do this, I think humanity is strong enough to get us through to the other side.

Yes, there will be heartbreak, but there will be lots of wins and laughter too, so keep going! Remember why you're here, remember what we have to do, and start treating this time—the most exciting time ever to be alive—as the adventure it truly is. A better world is possible, and we're already on our way.

All you have to do is get onboard and continue to show up for every win, every discovery, every exciting turn along the way.

And don't give up. Whatever you do, just don't give up.

CHAPTER 72
Paradise on Earth

The one thing we have to remember is that what we're trying to achieve is not a "one and done" thing. We won't find a climate-just world overnight, nor is this a battle to be won. The work we're here to do, to cocreate a different and (hopefully) much better world, will take decades. For many of us, it'll probably be our whole lifetime—*this is the journey of our lives!*

Daunting? I know it can be. But it's also quite exciting, isn't it? It means we are here to both witness and partake in the biggest shift humankind has ever gone through, a challenge I believe with conviction we can achieve if we put all our curious and courageous minds to the task. We are capable of so much—so, so much! As a species, we are truly extraordinary, and as soon as we recognize how much greatness we have to give this beautiful ecosystem called Earth, we will make miracles.

Let's remember to shift the narrative of our stories. We are not the evil termite to get rid of—we are meant to give and provide as much as we get to thrive. Living a new way will not mean sacrificing what we love, but choosing even more of what truly gives meaning, supports life. It simply means taking a second to pause and reflect on who we are and the world we want to create moving forward, for ourselves and all other species. We are powerful. With our power, we have created a lot of pain and destruction, but with that same power, we can create beauty, unity, abundance, and love. Let's start telling that story. Let our species be the conduit for paradise on Earth.

Gritted teeth and weapons will not save us this time—love, curiosity, courage, compassion, and excitement will. Whenever you find yourself frustrated by the lack of results or a fear that we won't make it, remember the journey. Remind yourself we're in it for the long haul. Even on the most depressing, cloudy, and gloomy days, days when the rain hits the windshields for hours on end and it's honestly quite pissy to be driving, we can and must just carry on. No matter what, we can't give up and turn around. We simply can't afford to.

There is so much to win, and when you think about the alternatives, so little to lose, so let's charge up our engines and find the courage to finally choose change.

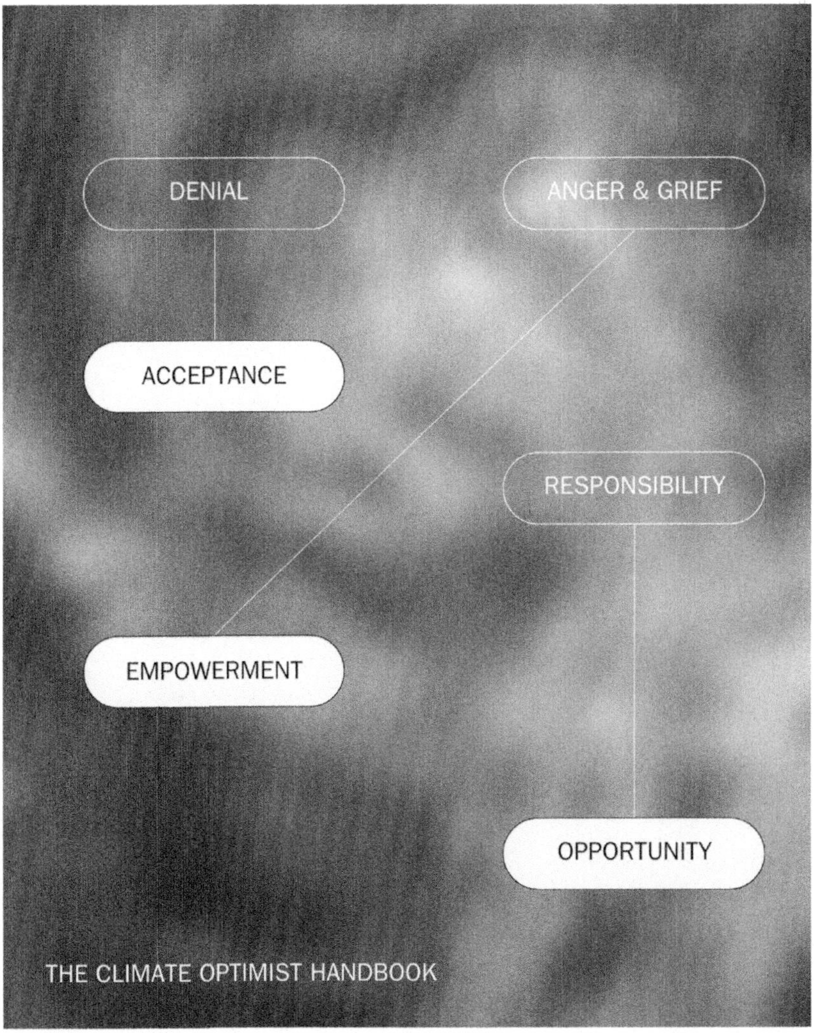

- DENIAL
- ANGER & GRIEF
- ACCEPTANCE
- RESPONSIBILITY
- EMPOWERMENT
- OPPORTUNITY

THE CLIMATE OPTIMIST HANDBOOK

PART SIX
CLIMATE OPTIMISM

CHAPTER 73
Things to Be Excited About

"It's not game over, it's game on!"
— *Paul Hawken, Author of* Drawdown *and* Regeneration:
Ending the Climate Crisis in One Generation

On the journey toward a different kind of world (hopefully one that's better than we can even imagine), we're already on our way. Much needs to happen, yes, but many solutions have already been discovered, gained excitement, and been put into play. We need to recognize the wins we're already experiencing and use them as fuel to keep moving forward. Again, we won't fix climate change overnight because it's technically not a problem to fix, but more of a journey we all must travel together. It will take years, even decades, to get this right, and for many of us, it could be our entire lifetimes. We might bring kids onto this journey who get to experience even more changes than what we're used to today. They will be kids of the future, for sure, but more than anything, they will be kids of change.

We're fueling that change right now, and we need to keep building upon momentum after momentum to allow for this seemingly unbelievable future to come true. But that will be a lot easier to do when we recognize the already unbelievable changes taking shape in our world today and how fascinatingly fast things can start to shift once we allow ourselves to believe them. Before sending you off on your journey (one I truly hope you choose to travel), I want to leave you with some examples of the change already in motion. Let me share some of the climate optimism that brings me excitement so that you can get excited about the same, or go out and seek your own!

Since this journey is ongoing and everchanging, I won't get too factual here. However, if something sparks your interest, I encourage you to act on that spark and explore the topic further. Maybe that's where *your* climate journey begins. Maybe this is a field that can bring you the meaning you've always longed for.

Things are rapidly moving in this space; therefore, new solutions and innovative ideas are constantly popping up around the world. So, continue to seek out climate optimism and stay especially focused on what is happening in your industry and community. Celebrate bold efforts, rejoice in every small win, and use others' courage and inspiration to start fueling some of your own. Who knows? Maybe your wild idea will be our next big move forward.

Here's what makes me excited (in the chaotic, euphoric mess that climate optimism exists in inside my brain):

- Regenerative farming practices that allow for soils to bounce back to life and for biodiversity to thrive
- Ocean restoration projects, including kelp farming, oyster farms, and plastic pollution cleanups
- Empowerment of Indigenous tribes where the people of the land get to have their voices heard and respected (we need more of this, but it's happening)
- The fact that some countries are beginning to give constitutional rights to nature
- Urban greening projects that focus on planting trees and green walls, inviting gardens on rooftops, and carbon-neutral transportation systems—efforts aimed to stabilize microclimates and make cities healthier and more livable for all
- Retrofitting of old buildings
- Solar fields and wind farms
- The development of cheap, efficient, green hydrogen fuel
- The further exploration of geothermal energy where we can use the core of the earth to heat and electrify the earth's surface
- Circular design models in fashion, car manufacturing, and other industries

- "C40 – Mayors in Action," a global network of mayors taking urgent action to confront the climate crisis
- Curbside composting systems, which enable composting for every citizen
- New innovative materials that can outface our use of plastics
- The reimagination of mobility to design the car of the future (or whatever future transportation will look like)
- Women empowerment and universal access to education (which has been scientifically proven to have a direct positive impact on climate change)[37]
- Recycled ocean plastics made into roads, buildings, computers, and other long-term product solutions
- New technologies aimed at capturing carbon in the atmosphere and making it into products like leather and plastics
- New building designs that enable carbon neutral or even carbon positive living
- Vertical farming; the emerging love for vintage and thrifted fashion; social media's ability to quickly share information that educates and inspires; the youth of today (OMG, young people, how are you so passionate and wise?)
- The fact that climate change is being talked about (let's talk more!)
- New kinds of mushrooms developing (nature is doing this all on its own) that can consume plastic (how awesome!)
- Nature's incredible resilience to bounce back.
- Our human will to survive and thrive.
- Intersectionality taking space in climate conversations and injustices of all forms being voiced and brought to the surface.
- Carbon capture projects
- Tree planting efforts from individuals and organizations; the emergence of stakeholder capitalism and conscious consumerism
- Grassroots movements making waves across the globe
- Crazy and bold people who dare to reimagine and come up with new ideas and solutions for materials, products, and infrastructure
- And lastly—you.

---

[37] Learn more about this at www.projectdrawdown.org

The fact that you've read this book makes me incredibly optimistic. Not only did you pick it up and explore what this whole climate optimism thing was about, but you finished it all the way through. So, as I'm writing this overlooking the Brooklyn skyline from where I popped down to eat my dinner by the East River in lower Manhattan, the late sun playing in the windows across the bay, I'm closing the book with a smile. I smile because I know you read it, and I know that a leader within you has been activated. I may never know what you do with that leadership, but I still get to smile because I don't have to know; all I have to do is trust—trust that you know your place in the world and the role you play in these important but exciting times we're living through.

As long as you dare to show up as an Optimist in Action, the world will continue to change. You know you're here to say yes to change, and I trust you will honor that mission. And together, we—you and I and everyone we continue to activate—get to be excited about the world we will make possible next.

The real challenge—and the real excitement—comes from breaking up old relationships with the world and what we think ought to be the way. That is especially true when it comes to our relationship with Mother Nature and the home she provides for us. To quote the "father of circular economy," William McDonough:

> *"Nature provides us with sources. It's our job to make them into resources, again and again, and again."*

I think we can do it, and I dare to believe you think we can too. Now, let's activate that change!

CHAPTER 74
Inspiration for Individual Action

Eager to get that optimism machine spinning, but don't know where to start? Here are some tips for individual actions you can start implementing today. Don't overwhelm yourself by trying every climate-positive action at once. Start in one category, create new habits and norms, and then move on to the next. And remember to have fun while you're at it!

These are just highlighted tips. If you want to dive deeper and learn more about a specific topic, there are tons of resources online.

FOOD
Composting

If I were to pick one favorite climate action, it would probably be composting. Not only is it easy once you get the hang of it, but it prevents your garbage from smelling as well! I like to think of the garbage can as a monster I do not want to feed, and I definitely do not want to feed it food. When food ends up in landfills, it gets compressed tightly without oxygen, a needed element for organic matter to naturally decompose. Instead, it ends up undergoing a slow, anaerobic decomposition process, which creates a byproduct called methane. Methane is a greenhouse gas that is more than twenty-five times as potent as carbon dioxide at trapping heat in the atmosphere,[38] and needless to say, we don't want to *fuel* climate change if we can avoid it.

Making sure your food waste gets composted instead of thrown in the trash is a simple way of limiting your negative footprint, but the benefits don't stop there. If you allow for food waste to compost, it gets to be broken down into fertilizer that can be added back to the soil, helping it become riches and more resilient. Richer soils are better at

---
38 United States Environmental Protection Agency (EPA). "Importance of Methane." https://www.epa.gov/gmi/importance-methane

holding water (preventing flooding), sequestering carbon from the atmosphere, and growing more nutrient rich foods. In other words, by simply composting your food, you are literally inserting yourself into the cycle of life. Not only are you *not* leaving a negative footprint, but you're actually creating a very positive one!

The ways to compost are many, so find one option that works for you, or start somewhere and advance over time. If you're a homeowner with a backyard, creating your own compost pile could be both fun and rewarding! It's not that hard; watch a quick YouTube video on how to create one and simply get going. The added win will be nutrients for your vegetables and flowers!

If you don't have a garden, there are many options for us city dwellers as well. Anything from worm bins to fancy kitchen gadgets (check out Lomi or Vitamix Food Cycler), to an option as simple as handing it off to someone else to do the job for you. If you wish to be a "Freezer Free-Rider," as I call it, you can simply store food scraps in your freezer or in a bin on the counter and drop it off with a local service. See if there's a municipal composting service in your city or perhaps a community garden or local farmer who would gladly take your "gold." A quick search online should give you an idea of what's available in your area.

Plant-Based

As previously mentioned, I am not trying to convince you that you have to go vegan. However, introducing plant-based options wherever you can is good for both you and the planet. Meat consumption is a leading cause of Amazon deforestation, and the amount of water needed to create one pound of beef is approximately 1,847 gallons (or approximately 700 liters for half a kilo of beef!)

For obvious environmental reasons, our meat consumption today is not sustainable, and with so many alternative protein sources at our disposal, we shouldn't have to consume so much. Not to mention the ethical concerns involving the animals raised in factory-farm conditions where a life is spent in crammed places with very little to live for, only to end up in a slaughterhouse way too early. We have a lot of power in turning the tide on climate change *and* taking a stance for how we'd like to see the world by being conscious with what we put on our plates, so don't take that choice too lightly. Besides, plant-based foods are both fun and delicious *and* leave you feeling so much better, most of the time. So, although you don't have to commit to any specific dietary lifestyle, introduce those veggies and grains to your plate whenever you can!

Order Purposefully (+ Bring a Container!)

As you already know, slowing down your life and mindset serves as a major climate action, and it serves you well when you're out eating as well. In the heat of the moment (and with a growling tummy), you might feel like you want to try everything on the menu. However, be honest with yourself, and take a moment to reflect on how much you'll actually eat. If you think you're ordering too much, reassess and think again. So much food is wasted simply because we can't finish what we ordered.

Zero Waste Tip: If you know you're one of those who can't help but order one dish too many, or if you're known for having a hard time finishing a plate, plan ahead and bring your own container. That way you can walk home with some delicious leftovers without having to ask the restaurant for a disposable container, which more often than not is either lined with chemicals or made from plastics, both of which are incredibly hard, if not impossible, to recycle.

Proper Planning

The same mindset goes for the food you cook and eat at home. Be smart with your grocery shopping by planning ahead, and try not to buy ingredients you probably won't use before they go bad. Plan what meals you want to eat for the week ahead, and bring a list to the store; that way you won't grab items that seem fun in the moment, but will most likely end up on a shelf in the fridge and going bad. By planning ahead, you can also check your fridge and pantry to see what you already have so you don't end up buying extra "just in case."

Also try to be aware of deals. "Three for two" can seem like a deal, but if you end up only eating one, then you're actually losing money, with food going to waste on the expense. Added mindfulness in the grocery store can take you far being a climate hero!

Leftovers

Put simply, break the stigma with leftovers, if you have any. Many times, food tastes *better* the next day, and if anything, it's an easy way to solve a problem—a hungry stomach, that is. If you're on the adventurous side, use leftovers to mix up a new stir-fry or put it in a blender for a delicious soup. You never know when a new masterpiece will emerge in your kitchen.

Food Saving Services + Apps

Approximately 17 percent of all food produced worldwide is wasted, according to a UN report.[39] In the US, make that 30-40 percent.[40] You'd be surprised to learn that much of that food is actually thrown away simply because it's not aesthetically pleasing enough. Fruits and vegetables that are too small, too big, or just a little weird-looking may never make it to any grocery store, and therefore, end up being wasted. What a shame!

Luckily, more and more services are available that collect this "ugly produce" and offer them to consumers, many times at a discounted price. If you're looking to save money, help eliminate food waste, and have groceries delivered to your doorstep, this might be the climate action for you! In the United States, there are services like Imperfect Foods, Hungry Harvest, and Misfits Market, but you can also check with your local farmer to see if they offer farm boxes or something similar. If you don't live in the US, see if there are any services with similar offerings in your area.

PACKAGING

Look for Aluminum, Paper, or Glass

Worldwide, only 9 percent of plastic is actually recycled.[41] That is because most plastic used in packaging is extremely hard to recycle and the value of recycled plastic is low. Therefore, try to avoid plastic packaging as best you can and always look for an alternative. Unlike plastic (which, if

---

39 Choi, Candice. "17% of Food Production Globally Wasted, UN Report Estimates." OPB. March 4, 2021. https://www.opb.org/article/2021/03/04/food-waste-global

40 U.S. Department of Agriculture, Food Waste FAQs: https://www.usda.gov/foodwaste/faqs

41 Agence France-Presse. "9 Percent of Plastic Worldwide Is Recycled, OECD Says." VOA News. February 22, 2022. https://www.voanews.com/a/percent-of-plastic-worldwide-is-recycled-oecd-says-/6455012.html

recyclable at all, can only be recycled once or twice before becoming so broken down it loses quality and value), aluminum is endlessly recyclable. That soda can of yours can become a new can again, and again, and again, contributing to a circular system worth striving for.

Paper and glass are also highly recyclable; just make sure to eliminate food scraps the best you can before adding them to the recycling bin. Also, read up on what standards apply at your local recycling center. Unfortunately, universal standards have yet to be created for how to properly recycle certain materials, so what is possible or not depends on the facility near you. Rule of thumb, though—don't throw something in the bin simply because you *want* it to be recyclable. This is called "wishcycling" and actually does more harm than good.

However, if we are to go back to choosing packaging, try to aim for aluminum (metal), paper, or glass over plastic. Glass comes with its own complications (it's heavy to ship), but at least it can return to the loop and become a new product.

Buy in Bulk

More and more stores offer bulk shopping these days, which is great! That way, you can buy your grains, oats, sugars, or nuts in your own bags (invest in a few produce bags if you wish) without having to grab products off the shelf with excessive packaging. It also allows you to buy whatever amount you want, which, in case you only need a little of something, also prevents you from wasting food as well.

Skip the Bag

This one you've probably heard one too many times, but it's as simple a climate action as there is! Say no to the plastic (or paper) bag at the store and bring your own—both stylish and efficient.

Bring Your Own...

...fill in the blank! The mindset of being independent from unnecessary, disposable packaging doesn't stop with bags. Eliminate paper cups and plastic lids at the coffee shop by bringing your own reusable cup. Say no to that plastic fork by making sure you always have your own ready to go in your bag. Bring a fancy to-go cup or a repurposed glass jar to that park party so when you're offered a drink in a plastic glass, you can kindly refuse and say you brought your own! (You'll be the coolest one at the party.)

I understand that carrying around a kitchen worth of stuff is not always convenient, but a little bit of a BYO mindset can go a long way.

Repurpose

Now this is a fun game! Instead of just throwing away packaging or a product once it's used up, tap into your creativity and see what purpose it could possibly serve next. I find old marinara sauce jars or peanut butter jars to be perfect storage for my bulk shops, beautifully displaying rice, seeds, and other grains in our kitchen.

My favorite repurpose win to date is probably one my husband came up with. When going through old stuff before a big move, he wondered what we could possibly do with two perfectly fine plastic covers from an old notebook. When I replied there was probably no use for them, he challenged me further and suggested that, maybe, they could be good cutting boards for our camping trips. I had to admit it was a brilliant idea, and lo and behold, in our AirBnb stay only a few weeks later, there were no good cutting boards to be found, and those plastic covers came in very handy!

Gift Wrapping

I get it; gifting someone a beautifully wrapped gift is sometimes worth more than the gift itself. But think about how much waste goes into something that will soon be ripped apart and thrown away. An estimated 4.6 million pounds of wrapping paper is produced in the United States each year, and of that, about 2.3 million pounds of it ends its life in landfills.[42] You may also be surprised to learn that most gift wrapping paper is actually not recyclable, and by throwing it in the recycling bin nonetheless, you can make an entire load unrecyclable.

I challenge you to be creative with your gift wrapping as well. If you're handing over a gift in person, you can use a scarf, blanket, or other beau-

---

42 Malone, Megan. "Holiday Tip: Don't Recycle Gift Wrap." Earth911.org. December 7, 2018. https://earth911.com/home-garden/holiday-tip-dont-recycle-gift-wrap/

tiful material you may find in your home and kindly ask for it back once it's been unwrapped. (You might want to clarify this to the receiver before the unwrapping begins.) I personally think old newspaper or even paper grocery bags make for beautiful wrapping as well, and you can decorate with natural ornaments like dried orange (so beautiful around Christmastime), a flower, or a twig. Let your imagination roam—the possibilities are fairly endless.

When you do receive a gift and wonder what to do with the waste, here's a trick to knowing whether or not you can recycle the paper. If the paper is metallic, has glitter on it, or has a texture to it, it is not recyclable. Try crunching the paper up into a ball. If it stays crunched, it's more than likely recyclable.

Shipping

Let's be honest, shipping isn't the best for eliminating either our carbon footprint or waste. However, there are ways to be smarter in this department too. For example, try to carefully open packaging when you receive it, and repurpose envelopes and boxes when you can. Look for more environmentally friendly packaging made from paper instead of plastic, or search for "plastic" options that are biodegradable. Try eliminating Styrofoam as much as you can (it's simply not recyclable), and choose biodegradable peanuts that will dissolve into white foam in front of your eyes. Add some water and poof, down the drain it goes—it's almost like magic!

EVERYDAY LIFE
Single-Use Items

We don't need them, we shouldn't want them, and it's time to create some new norms! Once you start to pay attention, you will notice these disposable, single-use items are everywhere, but that only means you have so many opportunities to make mindful choices. Make it a habit to bring your own bag, container, and utensils, and start saying "no" to disposables whenever you can. You will feel so empowered with every action, and remember, you are also planting seeds in the minds around you!

"Do I need this?"

How often do we buy something out of habit or because we think we need it, but we actually don't? How much of that perceived "need" is planted by commercials and other societal influences? Once you begin to slow down and invite more intention into your life, you will notice that more often than not, you don't need that coffee/chocolate bar/sweater/new TV/_____ (fill in the blank.) That in fact, you could be just as good if not better without it. (And your wallet will be better too!)

With one simple question—"Do I actually need this?"—you invite in a conscious decision to be made. That, in itself, is empowering. And if the answer happens to be "No, not really," then save that coffee for later when you can make yourself a cup at home and save one more paper cup from ending up in landfills.

Beauty + Hygiene

All it takes is one look into your bathroom cabinet or a walk down any beauty aisle to realize how much waste is included in our beauty routines. From packaging that can't be recycled (due to lots of plastics and mixed materials) to toxic ingredients that are neither good for the planet nor you and your skin, this department can feel like a jungle to navigate. Where do you even start?

Luckily, many sustainable and non-toxic alternatives are coming to market, and with a little bit of research, I'm sure you can find an alternative that better fits your values when it's time to fill up. Don't replace all your products all at once (this is not very sustainable), but use up what you have and slowly start replacing your beauty routine, one product at a time.

Repair + Pass On

Whatever it is, whether it be a chair, phone, or pair of socks, it's worth adding some love to the things in your life to make sure they last as long as they can. Mend a hole, repair what's broken, and pass on what you no longer wish to keep. Remember that one man's trash is another man's treasure!

HOME
Second-Hand/Antiques
Buying used is both fun and sustainable, and it usually comes with a better price tag as well. Hit up your local antique store or find a used marketplace online to scout for new furniture before buying something new.

Mattress
You might be surprised to learn that a lot of the furniture in your home could be emitting toxic chemicals through off-gassing. (Off-gassing happens when non-natural materials off-gas toxic chemicals that your body absorbs through your skin.) Your mattress is no different. Since you spend such a big amount of your life in your bed, choosing a natural mattress that doesn't off-gas could be a critical choice, both for your wellbeing and the environment.

Linen/Bedding
Off-gassing can also happen with the sheets you wrap yourself in at night. If you tend to sweat a lot at night, it could have more to do with your sheets than the room temperature. If you're wrapped in toxic chemicals, your skin will try to fight those toxins by sweating. A natural material like linen or organic cotton, therefore, is a much better option both for you and the environment.

Cleaning Products
Continuing down the same path, let's have a look at what's in your cleaning products. Products that you, by the way, spray across your entire home! If you wish for a safe and healthy environment, look for natural and zero-waste cleaning products.

Similarly, many cleaning products come in heavy excessive packaging, creating both plastic pollution and adding heavy shipping miles to the carbon balance. New companies are offering tablets that can be dissolved in water instead of buying the full product in the store. The products are just as good, but much better for the environment. Examples include laundry detergent in the form of sheets, and dish tablets that aren't wrapped in plastic. Look for these better solutions and make a big stance for the environment!

YOUR CLOSET

Perhaps no industry needs a serious makeover more than the fashion industry. It is responsible for up to 10 percent of global carbon dioxide output every year as well as a fifth of the 300 million tons of plastic produced globally.[43] It's time to rethink how we wear our clothes. Luckily, many brands are starting to approach this issue by upcycling materials or redesigning already existing garments. At the same time, new clothes are made without material blends that are hard to recycle (like polyester and nylon), and new technologies are evolving that enable those already circulating mixed materials to find a second life. Here are some tips on how you can help fuel this transition and invite circularity into your closet and outfits.

Love + Repeat

Nothing speaks trends more than fashion itself, so let's create some new trends and make it cool again to re-wear the same clothes. The idea that we need a new outfit for every party is silly to begin with and an outdated trend worth a fixer-up. The fastest, cheapest, and simplest way to become conscious with your fashion is simply to love what you already have and dare to repeat that outfit you love. Wear it to every party if you so wish. Get people talking and start shifting trends!

Clothing Swap Parties

How many items in your closet have you (hand on your heart) not worn in a very long time, if not at all? It's not your fault, so drop the shame. Living in a fast-fashion world that promotes the need for always buying new things makes it easy for the most conscious shopper to make an impulsive buy from time to time. When you recognize you can probably do a cleanse, what better way to create circularity out of that never-worn sweater than by hosting a clothing swap with your friends? Pick a date, send out an invite, and instruct your friends to bring clothes they're no

---

[43] Dottle, Rachel and Jackie Gu. "The Global Glut of Clothing Is an Environmental Crisis." *Bloomberg*. February 23, 2022. https://www.bloomberg.com/graphics/2022-fashion-industry-environmental-impact/

longer wearing. Add some snacks and drinks to the event and you can sip away while shopping each other's clothes—*for free!* Can you think of a better friends date?

Second Hand/Vintage

Enough clothing is already circulating the surface of this Earth to last us for a lifetime. So, instead of always buying something new, make it a habit to visit the local secondhand shops first. You might be amazed by the gems you find and the deals you'll get. Circularity is both fun and stylish.

Renting

Clothing rental has taken a new turn in recent years with both bigger and smaller companies offering outfits for renting. If you're someone who likes to have new outfits all the time, a weekly or monthly rental subscription could be for you. It's also a great option for special occasions when you want something a little nicer but aren't willing to spend the money, or when you can't justify buying something you might only wear once or twice.

Sustainable & Ethical

With shifting trends and increasingly conscious consumer habits, the supply of sustainable and ethical brands are stepping up to meet the growing demand. Sustainable fashion made from natural materials, with manufacturing processes using less energy and water; clothing that is repurposed and finding second life; and garments that are made on demand to eliminate deadstock and other waste, are slowly taking up space across the industry. Look for smaller brands that are sustainable to the core or support bigger brands' collections that are on the right path. Just be aware of greenwashing (when an organization spends more time and money on marketing itself as environmentally friendly than on actually minimizing its environmental impact) and always bring a critical eye. Sometime a product is not as sustainable as the company makes it sound, so a bit of consumer awareness can be helpful.

Nothing can be completely sustainable if it's also not sustainable for the people working the supply chain. Therefore, looking for ethical practices that care for garment workers, their families, and cultures is a huge part of the sustainable fashion journey.

Two great resources for knowing what brands to support are the websites Remake and Good On You. They both offer brand directories and reports.[44]

Conscious Washing

Last but not least, you have a tremendous opportunity to lower your clothing footprint by becoming more conscious about how you wash your clothes. Not only is the dryer one of the most energy-sucking machines in your household,[45] but drying your clothes in a machine also emits a lot of microplastics.[46] Therefore, simply hang-drying your clothes serves as a mega opportunity to be a climate hero in your day-to-day life.

When it comes to washing your clothes, wash on low temperatures as often as you can and fill the machine before you run it. Be mindful about how often you wash your clothes and whether, perhaps, a few hours of airing out would do the trick! When it comes to most athletic wear, it's even worse since most of those materials are synthetic, so they shed microplastics (think polyester and other performance materials). Invest in a Guppy Bag to wash your clothes in; it will catch those gnarly microplastics and prevent them from entering the ground water.

Lastly, look for alternatives to laundry detergent. I personally love the different options for detergent sheets. They come in paper packaging, and all you have to do is rip a sheet apart and let it dissolve into the water. No plastic waste involved and no extra carbon footprint from shipping heavy liquids to the store.

---

44 Remake Brand Directory: https://directory.remake.world/#/ Good On You Brand Directory: https://directory.goodonyou.eco/

45 Upton, John. "Your Clothes Dryer Is a Huge Energy Waster." Grist. June 12, 2014. https://grist.org/climate-energy/your-clothes-dryer-is-a-huge-energy-waster/

46 Kart, Jeff. "Your Clothes Dryer Is a Microfiber Exhaust Pipe." Forbes. October 8, 2020. https://www.forbes.com/sites/jeffkart/2020/10/08/your-clothes-dryer-is-a-microfiber-exhaust-pipe/?sh=6c-cbca4a3c1

YOUR MONEY
Your Bank

Believe it or not, when you put money in a bank, it's not just sitting there and waiting for you to come collect it years later. The bank puts that money to work; it goes out into the world to fund all sorts of projects. So while you're at home recycling and putting solar on your rooftop, your money could be out there funding projects that wipe all your efforts clean and then some.

Ninety-seven cents of every dollar that you put in your bank goes right back out into the world to fund stuff.[47] That is great because money should be put to work, and often it's funding really good projects, like roads, renewable energy projects, ground-breaking technology, etc. However, you might be upset to learn that trillions of dollars—your money—are fueled into environmentally damaging operations, like arctic drilling, fracking, tobacco, and oil.

Each year, the Rainforest Action Network comes out with a report called *Banking on Climate Chaos* where it exposes all the banks and how much money they're funding into various fossil fuel projects. Go to BankingOnClimateChaos.Org to find this year's report and learn the worst that banks are doing right now. If you're not happy with how your bank is doing, consider switching to a better one. Smaller banks like credit unions are usually a lot better, but you can also talk to the bank to ask what projects it is currently investing in, and what its outlook is on the future. Understand that where you put your money matters, so choose your bank wisely!

Your Investments

While banks can invest in dirty or clean projects, so can you, and there may be no better way to help fuel the transition into a better world than to help fund the ideas and visions that will take us there. Not only can you sleep well at night knowing your money is out doing good in the world,

---

[47] Coppola, Frances. "How Bank Lending Really Creates Money, and Why the Magic Money Tree Is Not Cost Free." *Forbes*. October 13, 2017. https://www.forbes.com/sites/francescoppola/2017/10/31/how-bank-lending-really-creates-money-and-why-the-magic-money-tree-is-not-cost-free/?sh=113020283073

but investing in a green and clean future is also a financially great idea!

Trends might vary and the opinions from people around you even more so, but if we're not investing in a world worth living in, are our investments worth anything at all? Who can cash out on a dying planet?

Check with your bank to see what investment options it offers or go with a platform that's solely focused on "doing good." If you're looking for more resources to get educated, the resources page at the website Women Power Our Planet is a good place to start.[48]

Voting with Your Dollar

The easiest way to make a positive change with your money is probably how you choose to spend it. Which products you buy in the grocery store and which brands you choose to support has a tremendous impact with unknown positive ripple effects. Not only will the stores look at those numbers for future purchase decisions, but you're also contributing to those shifting trends and cultures by making it okay to buy certain things. Just take a walk down any grocery store aisle today and you will see plentiful plant-based milk options and just as many alternatives for meats and other traditional products. It didn't use to be this way. Trends and people's spending habits gave way for a whole new marketplace!

Giving Back

Philanthropy work is not for everyone, but there is something to be said about gifting your hard-earned money to a cause you care about. It also serves as a great opportunity to do good in places where you can't be physically active but can still help support people who are passionately working to make this world a better place. I like to donate automatically to a few philanthropic organizations on a monthly basis. When I receive that "thank you for your donation" note in my inbox, it brings a smile to my face, and I didn't have to put any effort whatsoever into creating that change. Money left my bank account and entered another.

Energy flowed from me into something good. Sometimes creating change can be streamlined and simple.

---

[48] Find resources for bank options as well as green investing at

YOUR VOICE

Never think *your* voice doesn't matter. It's the most powerful tool you have. Not only can you spread awareness, curiosity, and optimism in the conversations you spark with the people around you, but by adding your voice to causes you believe in, you can help propel the shift into a different world.

Sign Petitions

It may seem like a silly act just to add your name or email address to a list, but signing petitions matters. Numbers speak for collective power, and that's how we trigger action. The people organizing these petitions need to prove other people care about the cause too, and they will use these lists of collected signatures to make their points heard. So, similar to donating to philanthropic causes, signing petitions is a great way to make change possible with very little action.

Engage With Your Local Representatives

Your politicians have been voted into position by you, the people; therefore, they are there to listen to your wishes and concerns. You may think the only way to have an impact on politics is every few years when there's an election, but that couldn't be more wrong. By calling into your local office and letting them know what changes you'd like to see in your town, or to express support for a new bill you'd like to see into law, you are making your voice heard. You would be surprised how much you can actually influence your local politics, *especially* if you've gathered a list of signatures to prove further support. The politicians are there for *you*, so never forget that!

Vote

If you are to start anywhere with your climate optimist actions, please start here. Educate yourself on whoever is running for office, what they stand for, and whether they are mapping out the future you'd like to see. Then *please* show up to vote whenever you can. We need systemic change in order to make a climate-just future take shape, and that means turning politics in the right direction.

HOMEOWNERS

If you own a home, you hold tremendous power to make a difference, all while creating a universe for yourself where more positive actions get to live and thrive. Your direct environment is a constant reminder of who you want to be and the world you want to see, so don't think that changing the world takes place far away from home. It happens right inside those walls where inspiration is sparked, hearts are changed, and insightful conversations are nurtured.

Here are some tips for how to make an eco-oasis out of your home!

Insulation

Nothing could be more energy efficient than making sure you have the right insulation. If you live in a slightly older home, make sure to check your insulation and update it where possible. Making sure no heat or cold air escapes your home saves both the environment and your money.

Solar Panels

The world of electricity, heating, and other energy consumption is ever-changing, with a growingly uncertain landscape on the horizon. Will energy prices continue to go up? You can avoid this uncertainty by joining the green energy space today. By installing solar panels on your roof or in your yard, you create energy independence while taking a big stance for climate change. Clean energy is the future, and luckily for all of us, solar panels are cheaper than we dared to predict only a few decades ago. In fact, the price of solar dropped 89 percent from 2009 to 2019—in only ten years![49]

When you look at the numbers, switching from dirty to clean energy only makes sense. Bill McKibben, writing for *The New Yorker*, states:

In 2021, reports read that "with current technology and in a subset of available locations we can capture at least 6,700 PWh p.a. [petawatt-hours per year] from solar and wind, which is more than 100 times global energy demand." And this will not require covering the globe with solar arrays: "The land required for solar panels alone to provide all global energy is

---

[49] Toussaint, Kristin. "The Price of Solar Electricity Has Dropped 89% in 10 Years." *Fast Company.* December 9, 2020. https://www.fastcompany.com/90583426/the-price-of-solar-electricity-has-dropped-89-in-10-years

450,000 km2, 0.3% of the global land area of 149 million km2. That is less than the land required for fossil fuels today, which in the US alone is 126,000 km2, 1.3% of the country." These are the kinds of numbers that reshape your understanding of the future.[50]

Say "yes" to solar and you partake in that future today!

Energy Supplier

If you don't have the ability to install your own solar panels, you can go green by changing your energy supplier to one that offers solar, wind, and other renewable sources of energy.

Retrofitting

You can retrofit your home in many ways besides updating your insulation and switching to solar panels. You can paint your roof white, which helps reflect sunlight, which helps cool down your home. You can install ceiling fans to be used instead of AC, helping cut energy use. Creating more wild areas in your yard is better for biodiversity and carbon sequestration than a lawn. You can also plant a green roof or install a system for repurposing greywater. All these are great options to climate-up your home, and there are many more. Stay curious and learn what would work best for you.

Composting

As I mentioned in the food section, composting is one of my favorite climate actions. If you have access to a backyard, you literally have the opportunity to create a pile of gold for no money. How you create a composting pile is up to you. You can either have an active compost pile that takes a little more work but creates fertilizer much faster, or go for a passive one that more or less takes care of itself. Just make sure you keep it far enough away from the house or locked up somehow so you don't invite unwanted guests (read critters and bears).

A quick search online should give you an ocean of articles and how-to videos to get started.

---

50 McKibben, Bill. "Renewable Energy Is Suddenly Startlingly Cheap." *The New Yorker.* April 28, 2021. https://www.newyorker.com/news/annals-of-a-warming-planet/renewable-energy-is-suddenly-startlingly-cheap

Grow Your Own Veggies

What could be more glorious than walking out to your own veggie garden on a late summer afternoon to harvest the produce for your dinner? The summer I started growing some of my own vegetables, I realized—not much! There is something so magical about being the harvester of your own love and hard work, and when it comes to shopping locally, it doesn't get much more local than this.

Growing your own vegetables is good for both your body and soul. It helps you ground and reconnect with Mother Nature, and it makes your appreciation for food so much stronger. If you're completely new to the green thumb game, start simple and go easy on yourself. It's okay to kill a plant or two in the learning process.

TRANSPORTATION

Your Car

If you find yourself in a life situation where living without a car feels impossible, do your absolute best to be mindful of its impact. Can you switch to an electric or hybrid? Could you maybe reorganize your family's schedule to enable more efficiency and car-pooling? Would you be okay with planning your trips better so you can shop for all the things you need in one trip instead of doing multiple runs to town? The fate of our world's ice sheets are literally in our hands, and they're melting faster than we can imagine.[51] In a reality like this, every mile matters, so make sure yours count!

Biking (or Other Forms of Carbon-Neutral Transportation)

Growing up in Sweden, I was brought up on a bike. And I still love biking! Despite the profanities that would occasionally slip out of my mouth in specifically nasty weather, biking provides a sense of freedom that a car or public transportation can never give you. Commuting to work or school via bike is not just good for your health, but it obviously leaves zero carbon footprint on the environment.

---

51 Carrington, Damian. "Fate of 'Sleeping Giant' East Antarctic Ice Sheet 'in Our Hands'—Study." *The Guardian*. August 10, 2022. https://www.theguardian.com/environment/2022/aug/10/east-antarctic-ice-sheet-in-our-hands-climate-action

Public Transportation

It goes without saying—the more of us who can share the footprint of one vehicle, the better. Choose public transportation whenever you can!

Air Travel

Now, this is a tricky one. In a favorable reality, all of us would stop flying today. Yet you know just as well as I do that won't happen. We shrunk our world with air travel, and it's a win we're not willing to give up anytime soon. However, you can make a huge difference by being conscious about how often you fly and for what reasons. Try to take fewer trips and stay for longer periods instead. Short weekend flights should be replaced with longer stays where you get to truly live and appreciate a new culture. Also try choosing destinations closer to home, and use other transportation like a train or boat when available. And when you do fly, use the feature that tells you which option comes with the lowest carbon footprint. (Google Flights now offers this per automation.)

If you're interested in knowing exactly how much polar ice your one flight melts, you can calculate that in seconds with Shame Plane by visiting www.shameplane.com.

Your Carbon Footprint + Offsetting

Like it or not, your existence comes with a carbon footprint. Your challenge (as it is mine) is to become aware of how big it is and find whatever ways possible to minimize it. That starts with calculating your carbon footprint. Luckily, many free resources help you to do so today. One worth checking out is Clever Carbon: www.clevercarbon.io.

Besides minimizing your footprint the best you can, you can also offset some or all of that carbon by purchasing *carbon credits*. That means you're investing in restoration and/or protection projects with the mission to balance the carbon levels in the atmosphere by either capturing new carbon or making sure carbon already captured doesn't get released. Carbon offset programs include everything from empowering Indigenous tribes in the Amazon rainforest that will safekeep ancient trees (that hold a lot of carbon) to kelp forestry and tree-planting projects. The options are many, as are the companies providing carbon credits, so if you want to start offsetting, do a little research and find one that fits you.

I will also mention that although carbon offsets are a great idea, they come with their fair share of complications. Measuring the accuracy of carbon stored or saved is still tricky, and clear guidelines have yet to be set for what can be counted as a carbon offset or not. There's also the mindset of, "As long as I keep offsetting, I'm cool," which doesn't really work. Carbon offsets are good for balancing things in the short term, but we still need to do all we can—and put all our creativity in play—to find solutions for a carbon-neutral, if not carbon-positive, world. In other words, don't just buy carbon credits. Look into all these other action opportunities to start living life as a true climate optimist!

CHAPTER 75
Terminology

There are a lot of terms in the climate and sustainability space, so it can be hard to keep track of them all. Here is a list of terms you might find useful.

CLIMATE OPTIMISM TERMS
As defined by me

Climate Optimism
Climate optimism is a mindset shift aimed to spark optimism, creativity, and solutions thinking while keeping a fact-based and awareness-driven approach to climate change and our future. It's about changing the narrative around climate change so we can act from courage and excitement, not fear.

Climate Optimist
A person practicing climate optimism. A climate optimist is someone who dares to believe in a better future, finds the courage to choose change, and starts practicing that change today by recognizing that we all have an opportunity to participate in the shift toward a better world.

Optimist in Action (OIA)
To be an Optimist in Action is to be the change you wish to see in the world, and in doing so, fuel your own optimism from within. As an Optimist in Action, you're not just hoping things will get better; you're proving to yourself they can. Even if it's just small things like saying no to a plastic lid at the coffee shop or taking the bike instead of the car, you begin to exercise this optimism regularly. Small actions build character and set

the framework for who you are, which will completely change how you view the world and what you believe is possible. By being an Optimist in Action, you start to believe in change.

Retruthing

This is a word coined by me with self-empowering and world-changing intentions. (Don't take this word too lightly.) Retruthing is a verb and the definition is:

> "The willingness to question what is and to let go of ideas, thoughts, and perceptions as they have lived in our heads up until now to create a sustainable and more compassionate world. Retruthing refers to one's ability to understand things and circumstances always change, and it is our duty as humans to adapt to those changes, to constantly find ourselves in new worlds, communities, and realities—new truths."

ABBREVIATIONS
PPM (Parts Per Million)

You may have heard this term a lot in climate conversations. PPMs speaks for the parts per million of carbon dioxide in the atmosphere. Prior to the Industrial Revolution, CO_2 levels were consistently around 280 ppm for almost 6,000 years of human civilization. We are currently at 416 ppm (as registered on August 27, 2022). The current global goal is to get ppm levels down to below 350 ppm, but other studies show that if we'd like to be able to live and thrive on this planet, we should probably aim for the level that has been proven to sustain human life before, which is below 300 ppm.[52]

IPCC (The International Panel on Climate Change)

The Intergovernmental Panel on Climate Change (IPCC) is the United Nations' body for assessing the science related to climate change. The IPCC prepares comprehensive Assessment Reports about the state of scientific,

---

[52] Read more about this in Climate Restoration: *The Only Future That Will Sustain the Human Race* by Peter Fiekowsky and Carole Douglis.

technical, and socio-economic knowledge on climate change, its impacts, its future risks, and the options for reducing the rate at which climate change is taking place.

COP (Conference of the Parties)

For nearly three decades, the UN has been bringing together almost every country on earth for global climate summits—called COPs (Conference of the Parties). In that time, climate change has gone from being a fringe issue to a global priority.

GHG (GreenHouse Gases)

GHG is short for "greenhouse gases" which refers to the different gases that contribute to our warming climate. Carbon dioxide (CO_2) and methane are the most common ones, but there are others, like nitrous oxide and water vapor. These greenhouse gases surround the Earth like a blanket in the atmosphere. That blanket prohibits sunrays from bounding back out into space, hence keeping that energy trapped in the atmosphere and heating up the planet. Very much like a greenhouse.

SDGs (Sustainable Development Goals)

Sustainable Development Goals (SDGs) are a collection of seventeen interlinked global goals designed to be a "shared blueprint for peace and prosperity for people and the planet, now and into the future." The United Nations General Assembly set up the SDGs in 2015 with the intention of achieving them by 2030.

ESG (Environmental, Social, and Governance)

According to Wikipedia, "Environmental, social, and corporate governance (ESG) is a framework designed to be integrated into an organization's strategy to create enterprise value by expanding the organizational objectives to include the identification, assessment, and management of sustainability-related risks and opportunities in respect to all organizational stakeholders (including but not limited to customers, suppliers, and employees) and the environment."[53] More and more companies are using

---

53 https://en.wikipedia.org/wiki/Environmental,_social,_and_corporate_governance

ESG goals to show market value to its stakeholders.

Environmental factors include the contribution a company or government makes to climate change through greenhouse gas emissions, along with waste management and energy efficiency.

Social factors include human rights, labor standards in the supply chain, any exposure to illegal child labor, and more routine issues such as adherence to workplace health and safety. A social score also rises if a company is well integrated with its local community and, therefore, has a "social license" to operate with consent.

Governance refers to a set of rules or principles defining rights, responsibilities, and expectations between different stakeholders in the governance of corporations.[54]

CLIMATE GOALS & SOLUTIONS

Net Zero

Net Zero is a goal you may have heard a lot about lately; it is claimed by both countries and companies. Put simply, the idea behind the net zero vision is to cut greenhouse gas emissions to as close to zero as possible, with any remaining emissions reabsorbed by, for example, the oceans, the forests, and human-made carbon capture projects. The goal is to find a balance where we can live with zero emissions. To keep global warming to no more than a 1.5°C increase this century—as called for in the Paris Agreement—emissions need to be reduced by 45 percent by 2030 and reach net zero by 2050.[55]

However, it's worth noting that reaching net zero may not be enough since the already existing greenhouse gases in the atmosphere will continue to destabilize our living conditions and make life here on Earth rather difficult. For more on this, and how we must strive for climate restoration, I recommend reading Peter Fiekowsky and Carole Douglis's book, *Climate Restoration: The Only Future That Will Sustain the Human Race*.

---

54 Robeco. "What Is ESG?" https://www.robeco.com/en/key-strengths/sustainable-investing/glossary/esg-definition.html

55 Un.org. "For a Livable Climate: Net-Zero Commitments Must Be Backed by Credible Action." https://www.un.org/en/climatechange/net-zero-coalition

Carbon Positive

For something—a product, building, food—to be carbon positive means the production of the product sequesters more carbon than it creates. This is usually achieved by eliminating carbon emissions while at the same time using practices that regenerate and restore, helping to balance out the carbon levels.

For example, foods grown regeneratively and either sold locally with little or no transportation, or transported with green energy and carbon neutrality in mind, can be carbon positive. A building that isn't just retrofitted or built to be as energy efficient as possible, but that also uses solar panels that create a surplus of energy, can be carbon positive as well. Similarly, new technologies that uses the carbon in the atmosphere to produce products like plastics, combined with thoughtful green production and transportation, can also end up carbon positive.

Carbon Capture & Storage (CCS)

The act of carbon capture is an industrial process that aims to reduce carbon emissions in a three-step process. First, you capture the carbon dioxide produced in an industrial activity, such as steel or cement making. Then you transport it somewhere it can be stored safely deep underground. Capture, transport, store.

Since reducing emissions is no longer enough, we must also do all we can to capture and store whatever carbon is already out in the open, as well as do our best to capture the carbon we will still produce in the decades to come. That is why there is quite a heavy focus on carbon capture right now.

There are also natural ways to work for carbon capture and storage. In fact, natural processes that draw carbon out of the atmosphere and sequester it in plants, soil, and rocks already exist.[56] Another carbon capture effort lies in better understanding how these natural processes can be enhanced. Those carbon capture processes include ecosystem capture (focused on enhancing earth's own ecosystems to sequester carbon regenerative farming and soil restoration, algae plantation, and forest restoration), geological restoration (accelerating the ability of rocks and minerals to

---

56 Yale planetary solutions project: https://planetarysolutions.yale.edu/center-natural-carbon-capture

lock up carbon), and using nature as a model in synthetic processes, where CO_2 can be converted into fuels or materials using both photosynthesis and natural geochemical processes as a model.

Climate Restoration

Climate restoration is the bold climate change goal, along with associated actions to restore the climate and bring CO_2 levels back below 300 ppm, which are the safe levels of ppm that humans have actually survived under for a long time.

Learn more about climate restoration efforts by reading *Climate Restoration: The Only Future That Will Sustain the Human Race* by Peter Fiekowsky and Carole Douglis.

Renewable Energy

Renewable energy is energy derived from natural sources that are replenished at a higher rate than they are consumed. Sunlight and wind, for example, are sources constantly being replenished.[57]

Renewable energy is the future in the sense that it doesn't require finding new resources to create it, nor does it pollute the environment in the use of its energy. However, there are still things to take into consideration, such as the fact that we need resources like lithium and cobalt to create batteries and solar panels, as well as steel to create wind turbines, which does not come without a footprint. Another challenge with renewed energy sources is that right now, they are fairly hard to store.

Green Hydrogen

Hydrogen is a colorless, odorless, and highly flammable gas that is used as fuel, especially in industrial processes. Green hydrogen is produced using only renewable resources, like solar and wind. Although only water is emitted when hydrogen is burned, creating it can be very carbon intensive. Green hydrogen differs from "grey" or "blue" hydrogen because no carbon is produced in its production.

Hydrogen is emerging as one of the leading options for storing energy

---

[57] Un.org. "What Is Renewable Energy?" https://www.un.org/en/climatechange/what-is-renewable-energy

from renewables and was featured in a number of emissions reduction pledges at the UN Climate Conference, COP26, as a means to decarbonize heavy industry, long haul freight, shipping, and aviation. Governments and industry have both acknowledged hydrogen as an important pillar of a net zero economy.[58]

I've learned that in Europe, hydrogen is sometimes explained as cheese: If solar energy is milk, green hydrogen is the cheese. This analogy comes from the fact that hydrogen can be stored for longer periods of time, while solar energy (at least at this point) can't. Green hydrogen is, therefore, seen as an important partner in the green energy transition.

Geoengineering

Geoengineering (or climate engineering) is the deliberate large-scale manipulation of an environmental process that affects the earth's climate in an attempt to counteract the effects of global warming. Examples include removing carbon dioxide from the atmosphere and different variations of solar geoengineering, like stratospheric aerosol injection and marine cloud brightening, with the attempts to bounce sunrays back into the atmosphere and slow down the warming of the planet.

Geothermal

Geothermal stems from the Greek words *geo* (earth) and *therme* (heat), and it's just what it sounds like—heat from the Earth. Geothermal energy is a renewable resource that taps into the Earth's core where this energy is continuously produced and is currently used to heat buildings and generate electricity.[59]

Stakeholder Economy

Stakeholder economy or stakeholder capitalism is a system in which corporations are oriented to serve the interests of all their stakeholders. These stakeholders include shareholders, but also employees, customers, and the

---

[58] World Economic Forum. "What Is Green Hydrogen and Why Do We Need It? An Expert Explains." December 21, 2021. https://www.weforum.org/agenda/2021/12/what-is-green-hydrogen-expert-explains-benefits/

[59] U.S. Energy Information Administration (EIA). "Geothermal Explained." https://www.eia.gov/energyexplained/geothermal

local community. Compared to shareholder capitalism, which only looks to the profit gain and interest of its shareholders (investors), stakeholder capitalism looks to benefit everyone and, therefore, provides a much more holistic view and approach. For example, a company following these values may make decisions that are not necessarily economically profitable in the short term, but those decisions look to sustain the environment and, hopefully, create economic growth as well in the long run.

Circular (Production/Economy)

The circular economy is a systems solution framework that tackles global challenges like climate change, biodiversity loss, waste, and pollution with the intention to keep materials, products, and services in circulation for as long possible. This usually starts at the design phase when materials are carefully selected and assembled in a way that the product can be taken apart at the end of its lifecycle and the materials repurposed into either a similar product or something else.

Cradle-to-Cradle (or c2c)

Similar to circular systems, the cradle-to-cradle concept is about designing and producing products in a way that at the end of their lives, they can be truly recycled or upcycled, or if not, safely returned to the earth. It's the opposite of the cradle-to-grave concept that the linear production system we are too well used to today speaks for, where products are produced and then disposed of.

COMMON CLIMATE TERMS

Sustainable/Sustainability

As defined by Wikipedia, "Sustainability is a societal goal that broadly aims for humans to safely co-exist on planet Earth over a long time. Specific definitions of sustainability are difficult to agree on and therefore vary in the literature and over time."[60] As the definition proposes, it can be hard to define sustainability today or to know what's actually sustainable. One definition I like to give the word sustainability is about caring:

---

60 https://en.wikipedia.org/wiki/Sustainability

Sustainability means to care. You care for yourself, other people, the things you own, and our precious planet. It can be confusing to know what is actually sustainable, but from an individual approach, it means you love and care for all you have. You recognize the world can only be truly sustainable if it is so for everyone. As soon as you start caring, sustainability will simply follow.

Zero Waste

Zero waste is a set of principles focused on waste prevention that encourages redesigning resource life cycles so that all products are reused. The goal is to avoid sending trash to landfills, incinerators, or the ocean.

Natural

You might see the term "natural" on many products but be aware—this claim doesn't actually mean much. A product can exist of questionable phthalates and other not-so-natural ingredients and still claim to be "natural." The only reason this is the case is compared to "organic," for example, there are no laws or regulations for this space. Therefore, it's a very unclear term that we should take with a grain of salt.

Organic

Organic means something has been grown or produced without the use of chemical fertilizers, pesticides, or other artificial agents. In the United States, organic means the product has been approved by USDA Organic.

Permaculture

Permaculture is the conscious design and maintenance of agricultural practices that promote diversity, stability, and resilience of natural ecosystems. It's an agricultural philosophy and practice that aims to mimic nature and integrate crops in the way nature designed for them to grow.

Vegan

A person who is vegan does not consume any animal products and typically does not wear or purchase other products that derive from animals, like clothing and beauty products. For a food or product to be vegan, no animals were used to create the product. However, just because something is vegan does not necessarily mean it was produced ethically or cruelty-free (although this is mostly the case.) For example, berries can be vegan but picked by people with inhumane pay and working conditions, in which case those berries are not ethical or cruelty-free. At the same time, a product can be vegan but not sustainable since vegan leather, for example, might be made from plastic and highly pollutant to the environment.

Agroforestry

As the word suggests, agroforestry is a combination of agriculture (agro) and forestry. Basically, it's a land use management system in which trees or shrubs are grown around or among crops or pastureland. That way, you can grow new crops without having to clearcut the area, and you can benefit from the trees in forms of shading and nutrients exchange. Agroforestry practices support agricultural production and help improve water quality and air quality, soil health, and wildlife habitat. These working trees can also grow fiber, food, and energy.[61]

Biodiversity

Biodiversity or biological diversity is the variety and variability of life on Earth. According to Wikipedia, "Biodiversity is a measure of variation at the genetic, species, and ecosystem level."[62] The Earth's health and balance is heavily reliant on biodiversity, which is why it's alarming that we're racing into extinction for so many of the Earth's species today.

The main cause of biodiversity loss is human activity caused by the demand for food, water, and natural resources that puts a heavy burden on the Earth's ecosystems. Biodiversity loss is also linked to climate change in a two-way equation. The changing climate is pushing many species toward extinction as temperatures become more extreme and the search for food more difficult, but biodiversity loss also fuels climate change, as the destruction of ecosystems undermines nature's ability to regulate green-

---

61 Benjamin, Jocelyn and Kate MacFarland. "Five Ways Agroforestry Can Grow Forest Products and Benefit Your Land, Your Pockets & Wildlife." USDA.gov. February 21, 2017: https://www.usda.gov/media/blog/2016/10/19/five-ways-agroforestry-can-grow-forest-products-and-benefit-your-land-your

62 https://en.wikipedia.org/wiki/Biodiversity

house gas emissions and protect against extreme weather, thus accelerating climate change and increasing vulnerability to it.[63]

That is why we must tackle climate change and biodiversity loss together.

Tipping Points

In climate science, a tipping point is "a critical threshold that, when crossed, leads to large and often irreversible changes in the climate system. If tipping points are crossed, they are likely to have severe impacts on human society."[64]

Examples of tipping points we're moving toward right now are the melting of the Greenland ice sheet, which contains enough water to raise global sea levels by more than twenty feet; the West Antarctic ice sheet (WAIS); the loss of the Amazon rainforest, which stores 200 billion tons of carbon and is home to millions of species of plants and wildlife; and the thawing permafrost, home to twice as much carbon dioxide than is already in the atmosphere. (Imagine if that gets released.)[65]

Greenwashing

Greenwashing is when an organization spends more time and money on marketing itself as environmentally friendly than on actually minimizing its environmental impact. There is more talk than action, so to speak. With a growing trend in sustainability, it can be tricky to know which companies are actually making strides and which ones are just trying to appear "good" in the eyes of consumers. However, we must stay aware of greenwashing and make sure to hold companies accountable for their words and actions.

Composting

Composting allows organic matter to break down and become nutrient-dense fertilizer or soil. When organic matter like food, garden waste,

---

63 European Commission. "Climate Change and Biodiversity Loss Should Be Tackled Together." https://ec.europa.eu/research-and-innovation/en/horizon-magazine/climate-change-and-biodiversity-loss-should-be-tackled-together

64 https://en.wikipedia.org/wiki/Tipping_points_in_the_climate_system

65 Cho, Renee. "How Close Are We to Climate Tipping Points?" Columbia Climate School. November 11, 2021. https://news.climate.columbia.edu/2021/11/11/how-close-are-we-to-climate-tipping-points/

and paper gets composted instead of ending up in a landfill, it does two really good things: 1) It creates nutrients to bring back to the soil, which makes the soil richer, healthier, and more resilient, and 2) It prohibits the creation of methane that takes place in a landfill when organic matter gets packed tightly without proper access to oxygen.

Compostable

Organic matter that can be composted is per definition compostable. However, something is only compostable if it ends up in a composting pile or municipality. Compostable matter will not automatically compost if thrown in the trash. As mentioned under "composting," compostable matter in landfills will end up undergoing a slower process called anaerobic composting, with the greenhouse gas methane being released as a byproduct.

Biodegradable

If something is biodegradable, it will break down naturally, without the help of a specific environment. However, it should be noted that everything is biodegradable at one point or another, and what we should focus on is not so much if something is biodegradable, but in what time that item will biodegrade. Some matter, like organic foods, can biodegrade within weeks, even days, whereas heavily processed materials like oil-based plastics can take up to hundreds of years.

More and more items are claiming to be biodegradable today. However, that doesn't necessarily mean the product is compostable. For example, plastic that is compostable is biodegradable, but not every plastic that is biodegradable is compostable. In order for biodegradable plastic to properly decompose, it needs to end up in a facility that reaches a high enough temperature for that material to break down. Typically, you will not reach this temperature in a backyard composting pile, so try leaving out those biodegradable plastic forks!

Also, before you reach for one of those compostable/biodegradable forks at the restaurant next time and think you're making a good choice, reflect on how likely it is that fork will end up in a composting facility. If it won't, it's just trash like everything else.

Recycling

Recycling is the action or process of converting waste into reusable material. The material in the product is broken down and remolded into something new. The recycling process differs depending on the material. Some materials (like steel and aluminum) have really good recycling qualities and can be recycled almost endlessly, while other materials (like plastics) lose quality for each recycling process and become less valuable over time. The recycling abilities for various materials also differ in different countries and regions, depending on the capacity of the facility of that region.

Repurposing

The fashion industry is jumping on the trend of upcycled and repurposed materials, where two pairs of old jeans may become a new denim jacket. The difference between recycling and repurposing is that repurposed material doesn't get broken down first, but is used in a state very close to its original form to create something new.

Upcycling

Upcycling means to reuse/recycle something in such a way as to create a product of higher quality or value than the original. For example, the use of recycled plastic bottles that are no longer useful to produce bags or shoes.

BYOB

You might be familiar with the "BYOB" terms as "Bring Your Own Beer" or "Bring Your Own Beverage." In the eco world, it may also stand for "Bring Your Own Bag."

FASHION

Sustainable Fashion

Sustainable fashion is an all-inclusive term describing products, processes, activities, and actors aiming to achieve a carbon-neutral fashion industry, built on equality, social justice, animal welfare, and ecological integrity.

Sustainable fashion concerns more than addressing fashion textiles or products.[66]

That said, sustainable fashion is not just about recycling plastic bottles or using organic cotton. Creating a fully sustainable fashion world means tackling the industry from multiple angles. If you're interested in fashion, it's a truly inspiring field to dive deeper into.

Ethical Fashion

Ethical fashion is fashion that aims to reduce the negative impact on people, animals, and the planet. While sustainable fashion is mainly focused on the environmental impact involved in producing and distributing fashion, ethical fashion looks to things like garment worker rights, culture appropriation, and animal rights.

Outfit Repeater

"Outfit repeater" is a trendy new word worth adding to your vocabulary! To be an outfit repeater simply means you wear your clothes on repeat and you're not afraid to show up in the same outfit to multiple occasions or parties. One of the biggest issues with the fashion industry is the pace at which we dispose of garments. By loving and reusing your items, you are automatically helping fuel the sustainable fashion movement.

Synthetic Fibers

Synthetic fibers are fibers made by humans through chemical synthesis, as opposed to natural fibers that are directly derived from living organisms, such as plants or fur from animals. Synthetic fibers are usually toxic both to humans and the environment. Examples of synthetic fibers are nylon, spandex, rayon, and polyester.

Natural Fibers

Opposed to synthetic fibers, natural fibers are made from living organisms, such as plants and animals. Natural fibers (if produced and treated in a non-toxic manner) fit into the cradle-to-cradle module where they can return safely to earth and biodegrade at the end of their life cycle. Natural

---

66 https://en.wikipedia.org/wiki/Sustainable_fashion

fibers are also much kinder to your skin, which is worth paying attention to. Your skin is your biggest organ, and whatever you wear on top of your skin (whether clothing or a beauty product) will be absorbed into your body. This is especially true when you're sweating and your pores are wide open. It's a shame that so much performative active wear today is made from synthetic materials.

PERSONAL & INTERPERSONAL TERMS

Eco-Grief

Eco-grief is the mourning of the loss of the natural world. Someone experiencing eco-grief feels sad and anxious at the knowledge that ecosystems are disappearing and species are going extinct.

Climate Anxiety

Climate anxiety is a state of anxiety brought about by worrying about climate change and what a continued heating planet could lead to for humanity and other species.

Climate Justice

According to Wikipedia, "Climate justice is a concept that addresses the just division, fair sharing, and equitable distribution of the benefits and burdens of climate change and responsibilities to deal with climate change."[67]

Climate justice means looking at the climate crisis and the solutions brought forth from a lens of justice for all, and with a special focus on peoples and communities currently facing the worst effects of climate change. Climate justice also aims to ensure that as we move forward into a sustainable future, we do so without further exploiting people, ecosystems, and cultures. Therefore, it's not just about finding the fastest, cheapest climate solution, but recognizing that short-sighted, poor decisions brought us to where we are now, and we must change our thinking moving forward.

---

[67] https://en.wikipedia.org/wiki/Climate_justice

Environmental Justice

Environmental justice focuses on changes in the environment more than changes in the climate. Of course, changes in our climate inevitably lead to changes in the environment, which is why working for justice in both is equally important and has to happen at the same time.

People working for environmental justice tend to focus on biodiversity in an area, preventing deforestation, laws against pollution, or other important work to keep our natural lands safe. Environmental justice is about protecting the environment around us by recognizing the natural world has a right to speak too, and that taking care of nature plays a vital role in our own wellbeing and survival.

Intersectional Environmentalism

According to Leah Thomas, climate activist and founder of Intersectional Environmentalist, "Intersectional environmentalism is an inclusive version of environmentalism that advocates for both the protection of people and the planet. It identifies the ways in which injustices happening to marginalised communities and the earth are interconnected. It brings injustices done to the most vulnerable communities, and the earth, to the forefront and does not minimise or silence social inequity. Intersectional environmentalism advocates for justice for people and the planet."[68]

Ecocide

Ecocide is a word for human impact on the planet that causes mass destruction on the environment.

Eco-Shaming

The Urban Dictionary defines eco-shame as "to shame another person for not respecting the environment." And that's exactly what it is—you make someone else feel bad because they're not doing anything for the environment, or if they are, they're not doing enough.

This "doing enough" part is worth pointing out because it's unfortunately a fairly common theme in the eco-conscious world. Perhaps you

---

[68] London Environmental Network. "Environmentalism in Action: Understanding Environmental Justice and Intersectionality." https://www.londonenvironment.net/environmentalism_in_action_environmental_justice_and_intersectionality

start taking sustainable action, and in the joy of that, you share your progress with the world around you, only to be met with feedback on how you could do better. Although it's important that we continue to educate ourselves and one another, eco-shaming is not productive. If you find yourself in the crossfire of eco-shaming or if you perhaps catch yourself eco-shaming someone else, bring yourself back to what you've learned in this book. Approach with love and good intentions and continue your empowering journey forward—no shame involved.

ACKNOWLEDGMENTS

I would like to acknowledge my patient and loving parents who always believed in me, no matter how crazy my ideas, and who continue to support and love me although I have chosen to live so far away. I wouldn't be where I am without them.

This book would not have been the same without my wonderful husband who never lets me have a thought without challenging it, and who has broadened my worldview in so many ways. It also takes a certain man to love and support you through thick and thin, especially when you're determined to change the world.

To all friends, mentors, and extended family in my life, I cannot thank you enough for your belief in and genuine excitement for my work. Whenever there was doubt, you were there, and you helped restore my belief in my message and encouraged me to keep going. If you have ever lent me a couch (or half your bed), invited me for a meal, or graciously taken me into your family, know that you are a huge part of this book and a big reason why I'm able to do the work I'm doing. You taught me that home truly is where your heart is.

To my brother, who's been my rock, my inspiration, and my closest ally all my life. It was always so important that you believed in me, and it continues to be important to this day.

To my editors, Tyler R. Tichelaar and Larry Alexander, you were wonderful to work with from day one, and you helped this book sing with the optimism it promises. I'd also like to acknowledge my coach and publisher, Susan Friedman, who offered more than I could ask for and served as the cheerful and solid rock I needed by my side on this journey.

Lastly, to Mother Earth, who continue to inspire, fuel, and guide me. Who would I really be without you? Would any of us be anything at all?

BIBLIOGRAPHY

Books

Eisenstein, Charles. *Climate: A New Story*. Berkeley, CA: North Atlantic Books, 2018.

Fiekowsky, Peter & Douglis, Carol. *Climate Restoration: The Only Future That Will Sustain the Human Race*. Irvington, NY: Rivertowns Books, 2022.

Foyer, Jonathan Safran. *Eating Animals*. New York, NY: Little, Brown & Company, 2009.

Gawdat, Mo. *Solve for Happy: Engineer Your Path to Joy*. New York, NY: Gallery Books, 2017.

Grant, Adam. *Think Again: The Power of Knowing What You Don't Know*. New York, NY: Penguin Random House, 2021

Gundry, Steven R. *The Longevity Paradox: How to Die Young at a Ripe Old Age*. New York, NY: Harper Wave, 2019.

Gundry, Steven R. *The Plant Paradox: The Hidden Dangers in "Healthy" Foods That Cause Disease and Weight Gain*. New York, NY: Harper Wave, 2017.

Hari, Johann. *Stolen Focus: Why You Can't Pay Attention—and How to Think Deeply Again*. New York, NY: Crown Publishing Group, 2022.

Hawken, Paul. *Drawdown: The Most Comprehensive Plan Ever Proposed to Reverse Global Warming*. New York, NY: Penguin Books, 2017.

Hayhoe, Katherine. *Saving Us: A Climate Scientist's Case for Hope and Healing in a Divided World*. New York, NY: Atria/One Signal Publishers, 2021.

Heath, Chip and Dan. *Switch: How to Change Things When Change Is Hard*. New York, NY: Currency, 2010.

Holthaus, Eric. *The Future Earth: A Radical Vision for What's Possible in the Age of Warming*. San Francisco, CA: HarperOne, 2020.

Karelas, Andreas. *Climate Courage: How Americans Are Bridging the Political Divide and Tackling Climate Change—A Bipartisan Citizens Guide*. Boston, MA: Beacon Press, 2020.

Rosling, Hans. *Factfulness: Ten Reasons We're Wrong About the World—and Why Things Are Better Than You Think*. London, Gr. Brit.: Sceptre, 2018.

Saad, Laila F. *Me and White Supremacy: Combat Racism, Change the World, and Become a Good Ancestor*. New York, NY: Sourcebooks, 2020.

Sharot, Tali. *Optimism Bias: A Tour of the Irrationally Positive Brain*. New York, NY: Vintage, 2011.

Sharot, Tali. *The Influential Mind: What the Brain Reveals About Our Power to Change Others*. New York, NY: Henry Holt & Co., 2017.

Stoknes, Per Espen. *What We Think About When We Try Not to Think About Global Warming: Toward a New Psychology of Climate Action*. Chelsea, VT: Chelsea Green Publishing, 2015.

Tickell, Josh. *Kiss the Ground: How the Food You Eat Can Reverse Climate Change, Heal Your Body & Ultimately Save Our World*. Miami, FL: Atria/Enliven Books, 2017.

Weber, Jack Adam. *Climate Cure: Heal Yourself and Heal the Planet*. Woodbury, MN: Llewellyn Publications, 2020.

Documentaries

An Inconvenient Truth. Dir. Davis Guggenheim. Lawrence Bender Productions and Participant Productions. Starring Al Gore. 2006.

Cowspiracy: The Sustainability Secret. Dir. Kip Andersen & Keegan Kuhn. A.U.M Films & First Spark Media. 2014.

Don't Look Up. Dir. Adam McKay. Hyperobject Industries. Starring Leonardo DiCaprio, Jennifer Lawrence, Meryl Streep, Cate Blanchett, Rob Morgan, Jonah Hill, Timothé Chalamet, and Arianda Grande. 2021.

Forks over Knives. Dir. Lee Fulkerson. Monica Beach Media. 2011.

Kiss the Ground. Dir. Joshua Tickell & Rebecca Harrell Tickell. Big Picture Ranch. Starring Woody Harrelson, Ray Archuleta, John Wich, Gisele Bündchen, and Ian Somerhalder. 2020.

Sacred Cow. Dir. Diana Rodgers. Nutrition Media. 2020.

The Need to Grow. Dir. Rob Herring and Ryan Wirick. Earth Conscious Films. Starring Rosario Dawson, Erik Cutter, and Douglas Gayeton. 2019.

Other Media Sources

Hey Change Podcast. https://www.theclimateoptimist.com/hey-change-podcast

Episodes mentioned:

E69: Bruce Lipton on Epigenetics, Living in the Matrix, and the Biology of Belief, Part 1

E70: Out Of The Matrix and Into a Climate Just Future—Bruce Lipton, Part 2

E86: Changing The Climate Conversation with Katharine Hayhoe

Joanna Macy Quote: *A Wild Love for the World.* Episode on *On Being Podcast* with Krista Tippett, April 25, 2019.

Arjuna Ardagh. The Radical Brilliance Project: https://radicalbrilliance.com/

Remake Brand Directory: https://directory.remake.world/#/

Good On You Brand Directory: https://directory.goodonyou.eco/

Women Power Our Planet – Money Resources: www.womenpowerourplanet.org/resources

Banking On Climate Chaos Report: https://www.bankingonclimatechaos.org/

ABOUT THE AUTHOR

Anne Therese Gennari started her journey as a climate optimist in her early twenties when she had a highly spiritual experience. Wiped out from anger and despair over the climate-changing world and the ignorance that fuels it, she collapsed on the floor in an extensive crying session that left her with one message: "You're here to be a climate optimist."

Ever since that day, Anne Therese has tried to figure out what being a climate optimist means. By learning and growing on this path, she has landed in a place of understanding of how real optimism is created and nurtured, as well as how to approach this important movement toward a climate-just future and world from a place of resilience, courage, and self-love.

Today, Anne Therese aims to share what she's learned in as many ways as possible. As a speaker and workshop host, she's delivered sessions to many acclaimed schools and companies. She's a TEDx speaker, podcaster, workshop host, and educator who has taught courses and lectured on climate optimism in various forms at numerous schools, including City College of New York, Columbia University, and the Fashion Institute of Technology. She has also worked collaboratively with companies like BMW and Tetra Pak on a collection of projects and has hosted motivational workshops for many more.

The Climate Optimist Handbook is the product of almost ten years of Anne Therese's passionate pursuit of knowledge and understanding about climate change. She hopes it will be a gateway for people everywhere to step into the empowered roles we ought to play during our time here on Earth.

Bring Climate Optimism to Your Next Event

Anne Therese Gennari would love to bring climate optimism to your next event. She's available for keynote speaking, workshop hosting, panel discussions, and more. It doesn't matter if you're a school, company, or NGO, climate optimism is for everyone!

A creative at heart, Anne Therese also loves to work collaboratively on projects and communications campaigns.

If you're looking for more inspiration, head over to The Climate Optimist website. There you will find the climate optimist master class, podcast, and other resources.

You can also find more information about Anne Therese and her offerings and contact her to find out how she can help you. Just visit:

www.TheClimateOptimist.com

Printed in Great Britain
by Amazon